Algebra
5–8

Written by
Hank Garcia

Editor: Carla Hamaguchi
Illustrator: Darcy Tom
Designer/Production: Moonhee Pak/Cari Helstrom
Cover Designer: Barbara Peterson
Art Director: Tom Cochrane
Project Director: Carolea Williams

Table of Contents

Introduction

Each book in the *Power Practice™* series contains over 100 ready-to-use activity pages to provide students with skill practice. The fun activities can be used to supplement and enhance what you are teaching in your classroom. Give an activity page to students as independent class work, or send the pages home as homework to reinforce skills taught in class. An answer key is provided for quick reference.

Each activity in *Algebra* has been tied to the NCTM standards as outlined below. The numbers of the related standards are listed next to each activity in the Table of Contents.

1. Number and Operations: Understand numbers, ways of representing numbers, relationships among numbers, and number systems.

2. Algebra: Understand patterns, relations, and functions. Represent and analyze mathematical situations and structures using algebraic symbols.

3. Geometry: Analyze characteristics and properties of two- and three-dimensional geometric shapes. Use visualization, spatial reasoning, and geometric modeling to solve problems.

4. Measurement: Understand measurable attributes of objects and the units, systems, and processes of measurement.

5. Data Analysis and Probability: Select and use appropriate statistical methods to analyze data. Develop and evaluate inferences and predictions that are based on data. Understand and apply basic concepts of probability.

6. Problem Solving: Solve problems that arise in mathematics and in other contexts. Apply and adapt a variety of appropriate strategies to solve problems.

7. Reasoning and Proof: Make and investigate mathematical conjectures. Develop and evaluate mathematical arguments and proofs.

8. Communication: Organize and consolidate mathematical thinking through communication. Use the language of mathematics to express mathematical ideas precisely.

9. Connections: Understand how mathematical ideas interconnect and build on one another to produce a coherent whole.

10. Representation: Create and use representations to organize, record, and communicate mathematical ideas.

Use these ready-to-go activities to "recharge" skill review and give students the power to succeed!

Real Number Line

The opposite of positive is negative. For example, 8 and ⁻8 are opposites. Opposites are the same distance from the origin, 0. Use the number lines below to locate the integers.

1 Plot 5 and its opposite.

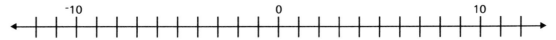

2 Plot ⁻11 and its opposite.

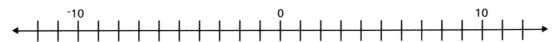

3 Plot 2.5 and its opposite.

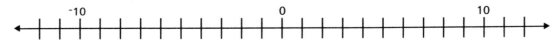

4 Plot ⁻$\frac{1}{2}$ and its opposite.

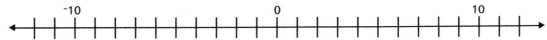

5 Match the given point on the number line with its corresponding letter to reveal a message.

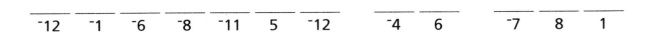

Adding Integers (Real Numbers)

$7 + (^-3) = 4$ $^-1.5 + 19 = 17.5$

1 $3 + (^-8) =$

2 $9 + (^-9) =$

3 $12 + (^-23) =$

4 $0 + (^-42) =$

5 $^-8 + (^-7) =$

6 $15 + (^-6) =$

7 $4 + (^-2) + (^-8) =$

8 $^-13 + (^-5) + 9 =$

9 $15 + (^-2.5) =$

10 $^-5 + 34 =$

11 $^-6 + 32.3 =$

12 $^-17 + (^-102) =$

13 $187 + 121 =$

14 $231 + (^-45) =$

15 $^-78 + 3.6 =$

16 $23 + (^-3.1) + 0.8 =$

Fill in the blank with **positive** or **negative**.

17 When a positive number and a negative number are added together, the sum is

_____ if the positive number has a greater absolute value.

18 If the negative number has a greater or absolute value, the answer is

_____.

Algebra © 2004 Creative Teaching Press

Name _____   Date _____

Subtracting Integers

$^-7 - 5 = ^-12$ $10.3 - (^-4) = 14.3$

1 $4 - 7 =$

2 $15 - 8 =$

3 $^-13 - 8 =$

4 $9 - (^-17) =$

5 $^-6.7 - (^-1.5) =$

6 $34 - (^-23) - 11 =$

7 $^-2 - 2 - (^-2) =$

8 $10 - (^-1.5) - 5 =$

9 $^-21 - 42 =$

10 $7 - 19 =$

11 $6.3 - (^-8.3) =$

12 $^-17 - 102 =$

13 $^-3 - 24 =$

14 $^-138 - 0 =$

15 $0 - 132 =$

16 $0 - (^-132) =$

Fill in the blank with **positive** or **negative**.

17   Adding a negative is like subtracting a _____.

18   Subtracting a negative is like adding a _____.

Time for an Operation

Other than multiplication, which operation is commutative and associative? Here's how to find out: Shade in only those boxes with the correct answer.

$5 + (^-3) = {}^-2$	$^-5 + {}^-3 = 8$	$^-4 - (^-7) = 2$	$3 - (^-3) = 0$
$11 - 7 = {}^-4$	$2 - (^-9) = 11$	$^-3 + (^-2) = 5$	$6 - (^-2) = {}^-8$
$10 - (^-3) = 13$	$^-4 + 23 = 19$	$6 + (^-12) = {}^-6$	$^-5 + (^-2) = 7$
$^-15 + 8 = 7$	$^-2 - (^-1) = {}^-1$	$20 - 32 = 12$	$0 - 15 = 15$
$22 - (^-1) = 21$	$8 - 31 = 23$	$^-9 + 24 + {}^-15$	$^-2 - (^-7) = {}^-5$

Algebra © 2004 Creative Teaching Press

Multiplying Integers

$$^-5 \times 4 = {}^-20 \qquad\qquad ^-8 \times (^-2) = 16$$

1 $6 \times (^-3) =$

2 $^-10 \, (4) =$

3 $(^-13) \, (6) =$

4 $15 \times 3 \, (^-2) =$

5 $^-4 \times (^-3) \times {}^-7 =$

6 $(^-12) \left(\dfrac{1}{2}\right) (^-3) =$

7 $(^-0.5) \, (18) =$

8 $^-2.5 \, (16) =$

9 $^-3 \, (15) \, (^-4) =$

10 $^-22 \times 22 =$

11 $^-106 \times 0 =$

12 $^-40 \times 0.1 =$

13 $(^-5) \, (4) \, (^-2) \, (0) =$

14 $^-8 \, (7) =$

Fill in the blank with **positive** or **negative**.

15 A positive times a positive is a _____.

16 A positive times a negative is a _____.

17 A negative times a positive is a _____.

18 A negative times a negative is a _____.

Dividing Integers

$$12 \div (^-4) = {}^-3 \qquad\qquad ^-8 \div (^-16) = \frac{1}{2}$$

1 $10 \div (^-5) =$

2 $^-16 \div (^-4) =$

3 $^-24 \div (^-2) =$

4 $0 \div (^-13) =$

5 $3.5 \div (^-2) =$

6 $75 \div (^-15) =$

7 $^-54 \div 18 =$

8 $-\dfrac{3}{4} \div 4 =$

9 $^-30 \div (^-6) =$

10 $1.2 \div 4 =$

11 $^-69.3 \div 3.3 =$

12 $^-51 \div (^-3) =$

13 $^-186 \div 3 =$

14 $4\dfrac{1}{3} \div 3\dfrac{1}{2} =$

Multiplication and division are "inverse operations." What other operations are inverse operations?

15 _____ and _____ are inverse operations.

16 An equation with an odd number of negative factors is _____.

17 An equation with an even number of negative factors is _____.

Algebra © 2004 Creative Teaching Press

Absolute Values

> The absolute value of an integer is the distance that integer is from 0.
> For example, $|5| = 5$, and $|^-5| = 5$. Also, $|2 - 6| = 4$.

1 $|^-3| =$

2 $|42| =$

3 $|7 - 9| =$

4 $^-|^-13| =$

5 $6 + |^-12| =$

6 $^-3 - |^-14| =$

7 $|9| + |^-34| =$

8 $|^-10| - |^-4| =$

9 $^-|25| - (^-|30|) =$

10 $^-|3| - 2 (^-4) =$

11 $|23| + |^-84| =$

12 $^-23 - |84| =$

13 $|0| =$

14 $^-|0| =$

15 $^-|2 \times 4| =$

16 $^-|32 \times 3| =$

17 What is the sum of any integer and its opposite? _____

Exponents

$$(^-3) \times (^-3) = (^-3)^2 = 9 \qquad\qquad 5^2 \times 5^2 = 5^{2+2} = 5^4 = 625$$

1 $4 \times 4 \times 4 \times 4 \times 4 =$

2 $(^-8)\ (^-8)\ (^-8) =$

3 $5^2 =$

4 $10^2 =$

5 $2^3 =$

6 $6 \times 6 \times 6^3 =$

7 $1^9 =$

8 $2^3 \times 2^4 =$

9 $(^-9)\ (^-9) =$

10 $21 \times 21 =$

11 $7^1 =$

12 $(^-1)\ (^-1)\ (^-1)\ (^-1)\ (^-1) =$

13 $0^5 =$

14 $(^-6)^3 =$

15 When a negative base (such as $^-5$) has an even exponent (such as ^{2}), is the result negative or positive? _____

16 When a negative base (such as $^-4$) has an odd exponent (such as ^{3}), is the result positive or negative? _____

Algebra © 2004 Creative Teaching Press

Homewrk Help

Al Gebra is failing his algebra course. Find the 10 errors in his homework and correct them.

1 $5 + (^-2) = 3$

2 $^-3 + (^-4) = 7$

3 $1.4 - (^-2.3) = 0.9$

4 $(5) (^-1) (^-2) = 10$

5 $(^-5) (^-5) (^-3) = ^-75$

6 $(^-6) (6) = ^-36$

7 $7.2 \div (^-9) = 0.8$

8 $2.3 (^-4.25) = ^-9.775$

9 $|^-45| + |^-6| = ^-51$

10 $|^-45| - |^-6| = 39$

11 $^-|^-45| + |6| = ^-51$

12 $(^-2) \times (^-2) \times (^-2) = 8$

13 $5^3 = 125$

14 $[12 - (^-6) \times (3)] \times (4 - 1) = ^-18$

15 $(7 - 3) \times [2 - (^-5 - 2)] = ^-4$

16 $|^-5| \times (3 - 4) = ^-25$

17 The integers $^-8$, $^-3$, $^-6$, 2, $^-1$ from least to greatest are written: $^-8$, $^-6$, $^-3$, 2, $^-1$

Order of Operations

P.E.M.D.A.S.

P = parentheses   **E** = exponents   **M** = multiplication
D = division   **A** = addition   **S** = subtraction

$3^2 + 6(5 - 2) = 27$ $(2 + 5) - 2^4 \times 1^5 = {}^-9$

1 $9(4 - 7) =$

2 ${}^-8 + (7^2 - 40) =$

3 $5 - 2 \times 6 =$

4 $18 + 8 \div 4 =$

5 $\dfrac{1}{2} \times 6 - 3 =$

6 $(8 - 4 \times 2) + 13 =$

7 $(7 + 4) \times [18 - ({}^-6) \times ({}^-3)] =$

8 $10 + 3 \div 3 =$

9 ${}^-3 - (15)({}^-4) =$

10 ${}^-(8 \times 7 - 2) =$

11 $(5^2 + 15) + ({}^-13) =$

12 $4[({}^-3)^2 + 1] =$

13 $(7 + 4) \times [18 - (6) \times ({}^-3)] =$

14 ${}^-8(7 - 7) =$

Note: Another way to remember the order of operations is **G.E.M.S.**
G = grouping symbols, **E** = exponents, **M** = multiplication or division, and **S** = subtraction or addition.

Algebra Awareness (Real Numbers)

❶ 3 + (⁻7) =

   a) 10   b) ⁻10
   c) ⁻4   d) 4

❺ |⁻31| + |⁻64| =

   a) 95   b) ⁻33
   c) 33   d) ⁻95

❷ 4.3 − (⁻2.5) =

   a) ⁻6.6   b) ⁻1.8
   c) 1.8   d) 6.8

❻ |8| − |⁻19| =

   a) 11   b) ⁻27
   c) ⁻11   d) 27

❸ (⁻7) (⁻3) (⁻2) =

   a) 48   b) ⁻42
   c) ⁻21   d) 42

❼ (⁻6) × (⁻6) × (⁻6) =

   a) (⁻6)^3 b) (⁻6^3)
   c) 36   d) 216

❹ 8.1 ÷ (⁻9) =

   a) 0.9   b) 9
   c) ⁻9   d) ⁻0.9

❽ [24 − (⁻10) × (⁻2)] × (5 − 2) =

   a) ⁻234   b) ⁻2
   c) 132   d) 12

❾ How would the integers 0, ⁻4, 8, 4 ⁻2 be written from the least to greatest?

   a) ⁻4, 0, 2, 4, 8   b) ⁻4, ⁻2, 4, 8
   c) ⁻4, ⁻2, 0, 4, 8   d) ⁻2, 0, 4, 8

Always Treat the Substitute Properly

The Substitution Principle: One expression may be replaced by another that has the same value. For example, if n = 3, then $^-4 + n = {}^-1$

Evaluate each expression if w = 4, x = 5, y = 1, and z = 7.

1 wx + y =

2 z (x – y) =

3 w (x + y) =

4 zx – y =

5 3wy + 2xz =

6 x (wz – 14y) =

7 3 (wy + 2xz) =

8 $^-4$ (wy + 3xz) =

9 xwz – 14y =

10 4wx – yz =

11 yxz – 5y =

12 (w + x) ÷ (x – w) =

13 [8 (x – y)] ÷ (8y – w) =

14 $^-2$wxy – 2z =

15 $^-2$wxy – ($^-2$z) =

16 2wxy – ($^-2$z) =

Combining Like Terms

Combining like terms is like going to the grocery store. All you need to remember is to put the apples with the apples and the bananas with the bananas.

For example, $2a + 4b + 3a - b + 2 = 5a + 3b + 2$

1 $7a + 3a =$

2 $^{-}7b + 5b + 3b =$

3 $7x + 6 - 4x =$

4 $3y + x + 2x =$

5 $3 - 7x + x =$

6 $^{-}3 + 5x + 7x + 5 =$

7 $3b + 4c - b =$

8 $5a - 4a + (^{-}3b) =$

9 $^{-}2x + 2y + 3x - y =$

10 $10f + 6e + 3e - 12f =$

11 $2xy + 3xy =$

12 $7ab - 6ab =$

13 $^{-}2xy + 2 - 4xy - (^{-}13) =$

14 $^{-}12abc + 2abc - 8 =$

15 $8ef + 2a - 5ef =$

16 $12uv - (^{-}6w) - 7uv =$

17 $^{-}7xy - 9 - (^{-}4xy) =$

18 $12x - x - (^{-}4x) =$

19 $^{-}2y^2 + 9y^2 =$

20 $2a^2 b^3 + 2a^3b^2 - 6a^2b^3 + 2a^2 =$

Equations with Addition and Subtraction

$$18 + n = 7$$
$$18 + (^-18) + n = 7 + (^-18)$$
$$n = ^-11$$

1 $n + 3 = 5$

2 $12 + n = ^-6$

3 $^-5 + n = 10$

4 $a - 30 = ^-23$

5 $b - 5.3 = 20$

6 $^-3 + 5x = ^-26$

7 $3 - y = 90$

8 $n - (^-13) = ^-54$

9 $^-4 - x = ^-32$

10 $8 + c = ^-3$

11 $n + 2.5 = 0$

12 $6 - (^-n) = ^-1$

13 $b - 32 = 78$

14 $^-41 + f = 56$

15 $9.3 - x = 0$

16 $n + \dfrac{1}{2} = 8$

17 $^-7 + c = ^-2$

18 $18 + d = 45$

19 $f - 12 = 1$

20 $2\dfrac{1}{3} - x = ^-5$

Algebra © 2004 Creative Teaching Press

Equations with Multiplication and Division

$$4n = 8$$
$$4n \div 4 = 8 \div 4$$
$$n = 2$$

$$\frac{1}{2}a = 3$$
$$\frac{1}{2}a \times 2 = 3 \times 2$$
$$a = 6$$

1 $3a = 60$

2 $^-15b = ^-30$

3 $f \div 6 = 4$

4 $\dfrac{n}{5} = 8$

5 $\dfrac{b}{^-9} = ^-2$

6 $^-5z = 40$

7 $\dfrac{x}{^-3} = 1$

8 $^-y = 8$

9 $^-x = ^-12$

10 $^-7 = n \div 5$

11 $^-\dfrac{1}{22}x = 0x$

12 $4.8n = 9.6$

13 $^-32 = 16a$

14 $14 = ^-2v$

15 $\dfrac{n}{6} = ^-\dfrac{1}{3}$

16 $\dfrac{b}{1.6} = 0.32$

17 You are designing a pool to go in your backyard. The length of the pool is 3 times its width. The length is 15 meters. What is the width?

Solving Multi-Step Equations

$$5n + 2 = 17$$
$$5n + 2 - 2 = 17 - 2$$
$$5n = 15$$
$$5n \div 5 = 15 \div 5$$
$$n = 3$$

1 $n + 6 = 38$

2 $^-2 + b = {}^-4$

3 $8f + 6 = 30$

4 $5 = 17 - 12a$

5 $6 (3 - x) = 54$

6 $\dfrac{1}{4}y + 3 = {}^-11$

7 $16 - 3x = {}^-11$

8 $\dfrac{5}{7}z + 1 = 11$

9 $\dfrac{2}{3}e + 7 = 17$

10 $^-6 = 3a - 12$

11 $^-5n - 3 = {}^-43$

12 $2.5b - 1 = 6.5$

13 $\dfrac{3x}{9} + 3 = 5$

14 $\dfrac{1}{2}n - 15 = {}^-11$

15 $\dfrac{n}{3} + 3 = {}^-1$

16 $\dfrac{1}{3}y + (^-7) = {}^-4$

Algebra © 2004 Creative Teaching Press

An "Old" Question

Where in the world did archeologists find the earliest math records? Solve the equations below to find out.

Answers

❶ $2x - 4 = 0$ _____

❷ $3n + 1 = 4$ _____

❸ $15 = 2y - (^-11)$ _____

❹ $100 = 4b$ _____

❺ $12c + 1 = 145$ _____

❻ $60 = 2h - 7 + 2h + 7$ _____

❼ $3a + 3 - 2a - 2 = 15$ _____

Answer Key

1	2	3	4	5	6	7	8	9	10	11	12	13
A	B	C	D	E	F	G	H	I	J	K	L	M

14	15	16	17	18	19	20	21	22	23	24	25	26
N	O	P	Q	R	S	T	U	V	W	X	Y	Z

The Distributive Property

$$3 \,(n + 1) = 15$$
$$3n + 3 = 15$$
$$3n + 3 - 3 = 15 - 3$$
$$3n = 12$$
$$3n \div 3 = 12 \div 3$$
$$n = 4$$

Simplify.

1 $4 \,(n + 1)$

2 $2 \,(a + 8)$

3 $^-7 \,(10 + 7f)$

4 $(2x + 3) \, 5$

5 $3 \,(n - 9)$

6 $5 \,(2z - 1)$

7 $(4 - 2k) \, 8$

8 $(15 - 4a) \, 5$

9 $6 \,(n + 1.2)$

10 $(n - 8.2) \, 5$

Solve.

11 $6 \,(n + 4) = 42$

12 $^-6 \,(x + 8) = 0$

13 $3 \,(^-4 + 2r) = ^-24$

14 $8 \,(^-7y - 6) = ^-160$

15 $(2a + 1.75) \, 2 = 7.5$

16 $(2.3 - 6n) \, \dfrac{1}{5} = 3.14$

Variables on Both Sides

$$5x - 7 = 21 + x$$
$$5x - 7 - x = 21 + x - x$$
$$4x - 7 = 21$$
$$4x - 7 + 7 = 21 + 7$$
$$4x = 28$$
$$4x \div 4 = 28 \div 4$$
$$x = 7$$

1 $^-2a - 2 = 6 - 6a$

2 $7x + 7 = 20 - 6x$

3 $4y + 8 = 6y + 2$

4 $7n + 10 = 98 - 4n$

5 $8 + 5z = {}^-2z - 48$

6 $^-7f - 10 = 8 - 4f$

7 $5n - 5 = {}^-7n + 139$

8 $2 + 5d = 2d + 29$

9 $3 + 6j = {}^-j + 59$

10 $4c + 9 = 5c + 18$

11 $^-2g - 2 = 5g + 75$

12 $7b - 2 = 4b - 2$

13 $^-4x + 10 = 2 + 4x$

14 $x + 1 = 7x - 53$

15 $^-5u + 5 = {}^-2u - 19$

16 $2 - 3h = 2h + 12$

17 $^-8 + 7x = {}^-7x + 48$

18 $6a + 7 = 7 + 3a$

Translating Word Problems

Algebra is like a written language. The trick with understanding a language is to translate it into something that makes sense. Translate the phrases below into the "algebraic language." Then solve the equations.

1 Seven times a number is 56. What is the number?

2 Fifty-one more than 9 times a number is 114. What is the number?

3 508 exceeds six times a number by 70. What is the number?

4 8 less than twice a number is 10. What is the number?

5 63 more than four-fifths of a number equals 111. What is the number?

6 A number decreased by ⁻16 is three times the opposite of the number. What is the number?

7 The sum of twice a number and 7 divided by 3 is 7. What is the number?

8 ⁻12 increased by half of a number is 28. What is the number?

Algebra © 2004 Creative Teaching Press

What's the Word with Equations?

Solve the word problems.

1 If the cost of 4 pears and 2 peaches equals $1.00 and the cost of 2 pears and 3 peaches equals $0.70, how much does each pear and each peach cost?

2 It takes Bob 5 hours to paint a house. It takes Jim 3 hours to paint the same house. How long will it take to paint the house if they work together?

3 The sum of twice a number and 13 is 47. Find the number.

4 Find a number which decreased by 10 is 4 times its opposite.

5 The sum of two numbers is 91. The larger number is 1 more than four times the smaller number. Find the numbers.

6 Four times a number plus three times the number is the same as 14 more than 5 times the number. Find the number.

Algebra © 2004 Creative Teaching Press

Consecutive Integers

When you count by ones from any number in the set of integers, you obtain consecutive integers. ⁻1, 0, and 1 are consecutive integers. Translate the phrases below to express consecutive integers. Then solve them. For example, the sum of three consecutive integers equals 18. $n + (n + 1) + (n + 2) = 18$.

1 The sum of two consecutive integers is 19. What are the integers?

2 The sum of two consecutive integers is ⁻23. What are the integers?

3 The sum of two consecutive integers is 49. What are the integers?

4 The sum of three consecutive integers is 69. What are the integers?

5 The sum of four consecutive integers is ⁻42. What are the integers?

6 The sum of four consecutive integers is ⁻106. What are the integers?

7 The product of two consecutive integers is 12. What are the integers?

8 The sum of two consecutive even integers is 170. What are the integers?

Un-Doing Equations

When you use the order of operations with real numbers, you are "doing" the equation. When you do an equation with variables, it is as if you are "un-doing" the equation. As such, instead of "un-doing" multiplication or division first, you "un-do" addition or subtraction first.

$$2x + 4 = 18$$
("un-do" + 4 first) $\quad 2x + 4 - 4 = 18 - 4$
(then "un-do" multiplication) $\quad 2x \div 2 = 14 \div 2$
$$x = 7$$

1 What would you do first with an equation like this $7(n - 2) = {}^-56 + 4n$?

    a) add
    b) consolidate variables on one side
    c) use distributive property
    d) divide

2 Then . . . $7n - 14 = {}^-56 + 4n$. What's next?

    a) add
    b) consolidate variables on one side
    c) use distributive property
    d) divide

3 Then . . . $3n - 14 = {}^-56$. What's next?

    a) add
    b) consolidate variables on one side
    c) use distributive property
    d) divide

4 Then . . . $3n = {}^-42$. What's next?

    a) add
    b) consolidate variables on one side
    c) use distributive property
    d) divide

    n = _____

Exponents and Variables

$5 \times n \times n \times n = 5n^3$ The square of $n + 3 = (n + 3)^2$

1 $(x) (x) (x) (x) =$

2 $(^-y) (^-y) (^-y) =$

3 $a \times a \times a \times a \times a \times a =$

4 $a \times a \times a \times b \times b =$

5 $2p \times 5p =$

6 $3 \times n (^-2) \times n =$

7 $9 \times s (^-6) \times s =$

8 $^-r \times r =$

9 $(^-d) (^-d) =$

10 $u \times u \times u \times v \times v \times v \times v =$

11 $2 \times k \times (^-4k) =$

12 $(^-n) (^-n) (^-n) (^-n) (^-n) =$

13 $(^-n) (^-n) (^-n) (^-n) =$

14 $2ab3 \times (^-2)ab^3 =$

Evaluate if $x = 3$ and $y = 2$.

15 $(xy)^2$

16 $(y - x)^3$

17 $x^3 + y^3$

18 $y - x^4$

Algebra © 2004 Creative Teaching Press

Volume Challenge

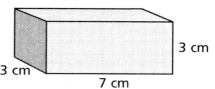 3 cm
3 cm
7 cm

The volume of a prism equals the base area times the height. So the volume of this rectangular prism equals 3 cm × 7 cm × 3 cm. Volume = 63 cm³

$V = Bh$

The volume of a triangular prism equals 1/2 times the base area times the height.

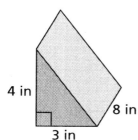
4 in
3 in
8 in

0.5 × 3 in × 8 in × 4 in = 48 in³

Find the volume of each figure.

1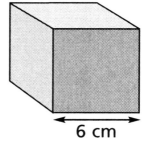
9 ft
2 ft
2 ft

2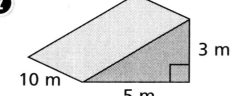
10 m
5 m
3 m

3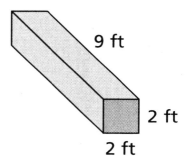
6 cm

4 Besides "n to the third power," how can n³ be read? _____

Algebra Awareness (Equations)

Evaluate if a = 2, b = ⁻3, and c = 4.

1 ab + c =

   a) 2   b) ⁻10   c) ⁻2   d) 10

Simplify.

2 4x + 3 + (⁻2x) − 4 =

   a) 2x + 1   b) 6x − 1

   c) 6x + 1   d) 2x − 1

3 r − 3.75 = 12.25

   a) 8.5   b) ⁻8.5

   c) 16   d) ⁻16

4 5n = 13

   a) 3   b) $\dfrac{13}{5}$

   c) $\dfrac{5}{13}$ d) $\dfrac{15}{3}$

5 21.5 = 7f + 4

   a) 2.5   b) 3.6

   c) 3   d) 55

6 18 (s − 5) = 18

   a) ⁻1   b) 1

   c) 6   d) ⁻6

7 5j − 2 = 8j + 1

   a) ⁻1   b) 1

   c) $-\dfrac{13}{3}$ d) $\dfrac{13}{3}$

8 The sum of two consectuive odd integers is 132. What are they?

   a) 64, 68   b) 66, 66   c) 65, 67   d) 61, 63

Add and Subtract Polynomials

$(2x^3 + 4x^2 - x) - (x^3 + 6x^2 - x) = 2x^3 + 4x^2 - x - x^3 - 6x^2 + x = x^3 - 2x^2$

A monomial is an expression that is a numeral, a variable, or the product of a numeral and one or more variables. The sum of monomials is a polynomial.

Add.

1 $2x^2 + 3$
   $\underline{4x^2 + 3}$

4 $5s^2 - t + u$
   $\underline{^-3s^2 - t + u}$

2 $^-2x^3 + 8x$
   $\underline{^-3x^3 - 7x}$

5 $^-5a^4 + b^2$
   $\underline{^-3a^4 - b^2}$

3 $n^5 - 3n$
   $\underline{^-2n^5 + 4n + 2}$

6 $^-6z^2 + y^2 + x$
   $\underline{5z^2 + y^2 + x}$

Add or subtract.

7 $(^-8x^2 - 3x^2 + 7) - (x^2 + 6x - 5)$

10 $(7x + 3) + (7x - 3)$

8 $(4x^3 - 3x + 4) + (9x^3 + 3x - 4)$

11 $(7x + 3) - (7x - 3)$

9 $(8x^2 - 5x^2 + 3) - (^-6x^2 + x + 1)$

12 $(^-3b^2 - 11b - 2) + (2a^2 - 7b + 1)$

Name _____ Date _____

Perimeter and Polynomials

You can use the addition of polynomials to find the perimeter of geometric figures. Add the sides of the figures below to find the perimeter.

1

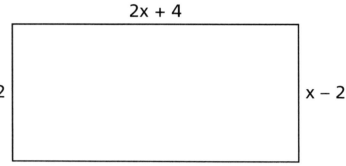

$2x + 4$

$x - 2$

$x - 2$

$2x + 4$

2

$6x^2 + 3$

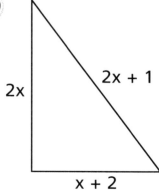

3

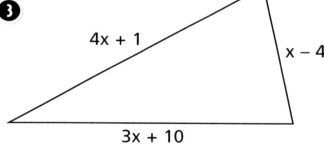

$4x + 1$

$x - 4$

$3x + 10$

4

$2x$

$2x + 1$

$x + 2$

Algebra © 2004 Creative Teaching Press

Multiplying Polynomials by Monomials

$2x (5x - 2) = 10x^2 - 4x$

$2x (x^3 - 2x^2 + 8)$
$2x (x^3) + 2x (^-2x^2) + 2x (8)$
$2x^4 - 4x^3 + 16x$

Simplify.

1 $9 (y^2 - 4)$

2 $5b (^-2b - 5b)$

3 $3f^2 (4f - 7)$

4 $^-w (4w^2 - 7w)$

5 $^-x (5x^2 + 2x - 3)$

6 $3a (2a^2 - 5a + 1)$

7 $2p^2 (p^3 - 2p^2 + 8p)$

8 $2m (2mn^3 + 3m^2 + 8n)$

9 $^-11b^2 (b^4 - 2b^2 - 5b)$

10 $^-a^3 (9a^3 - 7a^2 + 2a - 1)$

11 $4x^2 (6x^3 - 2xy + 2y^2)$

12 $^-3n^2 (6n^3 - 2n^2 - 2)$

Solve.

13 $2 (x + 4) - 1 = 9$

14 $6 (s - 2) + 3 (s + 1) = 9$

Multiplying Polynomials

Another way to multiply polynomials is to multiply vertically, as you would do with real numbers. This is helpful when multiplying polynomials. For example:

$$2x + 3$$
$$\times \quad x + 2$$
$$\overline{4x + 6}$$
$$2x^2 + 3x$$
$$\overline{2x^2 + 7x + 6}$$

First, multiply 2 by $(2x + 3)$.
Then, multiply x by $(2x + 3)$.
Add.

The trick is to make sure that every part of the first polynomial is multiplied by every part of the second polynomial.

1 $(2x - 1)(x + 3)$

8 $(p^2 - 2p + 8)(p + 5)$

2 $(3r + 4)(r - 5)$

9 $(6y^2 - y + 4)(9y - 1)$

3 $(j + 3)(j - 3)$

10 $(p^2 - 2p + 8)(p + 5)$

4 $(2y^2 + 4)(2y + 1)$

11 $(3n^2 - 5n + 8)(n - 10)$

5 $(8n^2 - 6)(n - 9)$

12 $(s - 3)(5s^2 - 2s + 4)$

6 $(x^2 - 4x - 7)(x + 5)$

13 $(5x^2 + 4x + 9)(x - 1)$

7 $(b - 1)(2b^2 - b + 3)$

14 $(r - 11)(4r^2 + 7r - 8)$

Distance, Rate, and Time

Distance = Rate × Time

Solve the word problems.

1 Danny drove to his friend's house at 48 mph. His friend lives 136 miles away. He drove home the next day at 60 mph. How many hours in total did Danny spend driving?

2 Emily drove to Tim's house at 40 mph. Tim's house is 48 miles away. Emily arrived at Tim's house at 11:23 a.m. What time did Emily leave her house?

3 At midnight, a jet left Seattle, Washington for St. Louis, Missouri, 2,100 miles away, flying at 500 mph. One hour later a high-speed jet left St. Louis for Seattle at 700 mph. At what time did they pass each other?

4 Two jets leave an airport at the same time, but travel in opposite directions. The first jet is traveling at 377 mph and the other at 275 mph. How long will it take for the jets to be 2,065 miles apart?

Name _____ Date _____

Homework Help

Our friend Al Gebra is improving in his algebra skills. However, he is still failing. Please help him with his homework. Find and correct his 7 mistakes.

1 (s) (s) (s) = s^3

2 ($^-$x) ($^-$x) ($^-$x) ($^-$x) = $^-x^4$

3 (7p + 8) (2p − 3) = 9p + 5

4 (4g + 2) − (6g − 4) = $^-$2g − 2

5 n (n^2 − 5n) = n^3 − $5n^2$

6 z ($4z^2$ + 3z) = $4z^3$ + $3z^2$

7 z (z^2 + 3z − 1) = $4z^3$ + 3z − z

8 (b − 5) (b + 2) = b^2 − 3b − 10

9 (5a + 3) (4a − 7) = $20a^2$ − 23a − 21

10 (3x + 1) (5x − 9) = $15x^2$ − 22y − 9

11 (3x − 2) (3x − 2) = $9x^2$ − 12x + 4

12 (6d + 4) (6d + 4) = $36d^2$ − 48d + 16

13 (2e − f) (2e − f) = $4e^2$ − 4ef + f^2

14 $(5x − 6)^2$ = $25x^2$ − 60x + 6

15 $(2n − 3)^2$ = $4n^2$ − 12n + 9

16 $(x − y)^2$ = x^2 − 2xy − y^2

Name _____ Date _____

Area and Polynomials

The area of a rectangle equals its length times its width (A = LW).
The area of a triangle is half of the area of the same product $A = \frac{1}{2} LW$

Find the area of each figure.

 1

2x – 2

x

 2

5x – 4

Find the area of each shaded triangle.

3

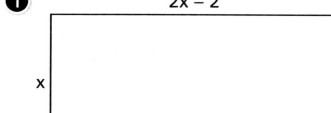

x – 8

2x – 7

4

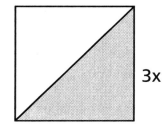

3x

Area and Perimeter Challenge

Add, subtract, or multiply to solve the problems.
For example:

Multiply $(x - 4)(2x + 2)$
                   $2x^2 - 6x - 8$

Multiply $(x - 8)(x - 3)$
                   $x^2 - 11x + 24$

Subtract $2x^2 - 6x - 8$
        $- \quad x^2 - 11x + 24$
             $x^2 + 5x - 32$

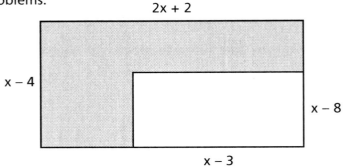

1 Find the area of the shaded region.

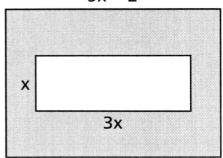

2 Find the total area of the shaded regions.

area = $x^2 + x$

area = $2x^2 + 4x + 2$

3 What is the area of this parallelogram?

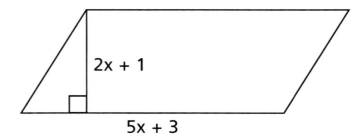

Algebra © 2004 Creative Teaching Press

FOILed

A handy acronym in multiplying binomials is F.O.I.L. (First, Outside, Inside, Last)

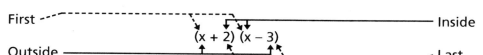

First   Inside

$(x + 2)\ (x - 3)$

Outside   Last

The result is often a trinomial. $x^2 - x - 6$

① $(x + 3)\ (x - 1)$

② $(x - 4)\ (x + 5)$

③ $(2x - 4)\ (3x + 4)$

④ $(7x + 4)\ (9x - 4)$

⑤ $(12x - 2)\ (3x + 1)$

⑥ $(5x + 4)\ (x - 1)$

⑦ $(7x - 2)\ (2x - 3)$

⑧ $(2x^2 + 4)\ (x^2 - 6)$

⑨ $(x^2 - 9)\ (5x^2 - 7)$

⑩ $(3x^2 + 3)\ (3x^2 + 3)$

⑪ What happens when you multiply two binomials like $(x + 1)$ and $(x - 1)$ together? Does it result in a trinomial? Why or why not?

⑫ Try another: $(2x^2 + 3)\ (2x^2 - 3)$

Squaring Binomials

A binomial which is squared results in a perfect square trinomial.

$(a + 2)^2$
$(a + 2)(a + 2)$
$a^2 + 4a + 4$

1 $(x - 1)^2$

6 $(9x + 4)^2$

2 $(3x + 1)^2$

7 $(2x - 7)^2$

3 $(3x + 4)^2$

8 $(x^2 + 6)^2$

4 $(x - 4)^2$

9 $(4x^3 - 2)^2$

5 $(2x - 2)^2$

10 $(8x^4 - 3)$

11 If you take the trinomial $(4x^2 + 8x + 4)$ and divide it by $(2x + 2)$, what will you get?

12 Try another: $(9x^2 - 6x + 1)$ divided by $(3x - 1)$

Algebra © 2004 Creative Teaching Press

Polynomial Equations

$$3(n + 3) - 7 = 8$$
$$3n + 9 - 7 = 8$$
$$3n + 2 = 8$$
$$3n = 6$$
$$n = 2$$

1 $2(a - 3) + 5 = 7$

8 $15 = 3(y - 1) + 2(4 - y)$

2 $0 = 3(1 - 2v) - 5(2 - v)$

9 $(x - 1)(x - 3) = (x + 2)(x - 5)$

3 $c(2 - 3c) + 3(c^2 - 4) = 0$

10 $(n - 3)(n + 7) - (n + 1)(n + 5) = 0$

4 $4(n - 7) - 2n(1 - 3n) = 6n^2$

11 $(2a - 5)(a - 4) + 2a(1 - a) = 0$

5 $b^2 - [4 - b(3 - b)] = 2(2b - 3)$

12 $(x - 1)(6x + 5) = (3x + 5)(2x - 3)$

6 $1 = 2x(x - 2) - [5 - 2x(1 - x)]$

13 $(x^2 - 4x + 5)^2$

7 $2(a - 3) + 5 = 7$

14 $(x + 5)^3$

Name _____ Date _____

What's the Word with Rectangles?

Solve the word problems.

1 A rectangle whose perimeter is 68 feet has a length that is 8 feet longer than its width. What is the area of the rectangle?

2 A rectangle and a square have the same area. The length of the rectangle is 48 meters more than two times its width. The length of a side of the square is 48 meters. The side of the square is 72 meters less than five times the width of the rectangle. What are the dimensions of the rectangle?

3 The width of a rectangle is half the size of its length. The perimeter is 27 feet. What are the dimensions of the rectangle?

4 Danny has measured the area of a square and is comparing it with that of a rectangle. The length of the rectangle is 54 inches more than the length of the square. The width of the rectangle is 8 inches less than fifteen times the width of the square. The area of the square is 25 square inches. What are the dimensions of the rectangle?

Algebra Awareness (Polynomials)

1 ($^-$s) ($^-$s) ($^-$s) =

    a) s^3 b) $^-s^3$ c) s^2 d) $^-(^-s^3)$

2 $(2a + 5) - (3a - 4) =$

    a) $^-a + 1$ b) $a + 9$ c) $^-a + 9$ d) $a + 1$

3 $^-x (3x^2 - 7x - 2) =$

    a) $^-3x^2 - 7x - 2$ b) $3x^3 + 7x^2$ c) $3x^3 + 7x^2$ d) $^-3x^3 + 7x^2 + 2x$

4 $(2n - 6) (5n + 4) =$

    a) $10n^2 - 20n$ b) $10n - 24$ c) $n^2 - 20n - 24$ d) $10n^2 - 22n - 24$

5 $(3x - 2)^2 =$

    a) $9x^2 - 12x + 4$ b) $9x - 12x + 4$ c) $9x^2 - 12x$ d) $^-9x^2 - 12x + 4$

6 $(r - 1) (r^2 + r - 2) =$

    a) $r^2 - 3r + 2$ b) $r^3 - 3r + 2$ c) $r^3 - 3r^3 + 2r$ d) $r^3 + 2r^2 - 3r + 2$

7 ($^-$s) ($^-$s) =

    a) s^2 b) $^-s^2$ c) s d) $^-(s^2)$

8 Find the area of the figure.

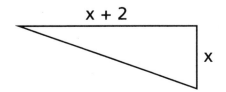

                                  a) $x^2 + x$ b) $^-x^2 + x$

                                    c) $x^2 - x$ d) $\dfrac{x^2}{2} + x$

What's the Word with More Equations?

Solve the word problems.

1 The sum of a number and 15 is equal to twice the number plus 5. What is the number?

2 In the first year of production, the senior class sells 1,572 tickets for its play. In the second year, it sells 1,753 tickets. In the third year, it sells 152 less tickets than in its second year. How many tickets are sold in all 3 years?

3 John reads at an average rate of 30 pages per hour, while Jeremy reads at an average rate of 40 pages per hour. If John starts reading a book at 4:30 p.m., and Jeremy begins reading the same book at 5:20 p.m., at what time will they be reading the same page?

4 Bill has twice as much money invested in stocks as in bonds. Stocks earn 8% interest per year and bonds 2% per year. If Bill earned a total of $225 from his stocks and bonds last year, how much money did he have invested in stocks?

Common Factors

To find common factors, first find the prime factorization. You can use a factor tree if you wish.

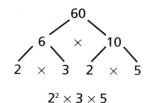

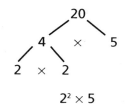

Since the factors of 60 are ②×②× 3 × 5 ...
... and since the factors of 20 are ②×②× 5 ...
... common factors multiply to yield the greatest common factor 4.

Find the greatest common factor for each set of numbers.

1 14, 42

2 24, 32

3 45, 52

4 42, 63

5 33, 99

6 49, 84

7 168, 420

8 121, 143

9 160, 180

10 180, 196

11 160, 196

12 12, 16, 18

13 20, 24, 32

14 36, 48, 60

Greatest Monomial Factor

You can find the greatest monomial factor as you did with greatest common factor. For example:

$24a^3b^2 = 2 \times 2 \times 2 \times 3 \times a \times a \times a \times b \times b$ and $18a^2b^4 = 2 \times 3 \times 3 \times a \times a \times b \times b \times b \times b$

Therefore, the greatest monomial factor is $2 \times 3 \times a \times a \times b \times b = 6a^2b^2$

Find the greatest monomial factor of the given monomials.

1 $8x, 4$

2 $24n, 12n$

3 $y^2, 13y$

4 $10c^2, 35c$

5 $8x^3, 3x^3$

6 $34n^2, 51n, 17$

7 $24a^3, 3a^2, 3a, 3$

8 $24a^3, 3a^2, a, 3$

Find the missing factor.

9 $24x^5y^3 = (6x^3y^2)\ (\ ?\ )$

10 $(^-4m)\ (\ ?\ ) = {}^-12m^3n^5$

11 $(3a^3b^2)\ (\ ?\ ) = {}^-18a^4b^5$

12 $(\ ?\ )\ (8c^6d^3) = 24c^{12}d^4$

13 $36x^8y^2 = (^-6x^3y^2)\ (\ ?\ )$

14 $(\ ?\ )\ (4s^5t^2) = {}^-48s^9t^3$

Algebra © 2004 Creative Teaching Press

Prime Time

Which positive integer is the smallest prime number? Here's how to find out. Shade in the boxes in which the term on the bottom is the greatest common factor (or GMF) of the terms on the top.

12, 18 4	24, 32 8	2m, 4 2	10n, 15n 5n
2, 4 4	4r, 16r r	42, 34 6	20a, 32a 4a
8k, 4 2	12j, 32j 2j	44x, 22x 22x	12, 18 4
52h, 48h 4	2xy, 7xy xy	28p, 14p 7p	81q, 72 9q
121w, 99w 11	8, 12, 24 4	4z, 6z, 8z 2z	6b, 8b, 24b 2b

Algebra © 2004 Creative Teaching Press

Name _____ Date _____

Difference of Squares

$$x^2 - 25 = (x + 5)(x - 5)$$
$$3b^2 - 27$$
$$3(b^2 - 9)$$
$$3(b + 3)(b - 3)$$

Factor.

1 $x^2 - 1$

2 $s^2 - 16$

3 $v^4 - 36$

4 $n^2 - 81$

5 $4d^2 - 4$

6 $25n^4 - 16$

7 $16n^2 - 49$

8 $m^2 - 25$

9 $9a^2 - 100$

10 $625x^4 - 1$

11 $s^4 - 81v^4$

12 $x^8 - y^4$

13 $81m^4 - 16n^4$

14 $1 - 121c^{16}$

Circular Spaces

The area of a circle equals π (radius)². A = πr²
Find the areas of the shaded regions. Here's an example with a square.

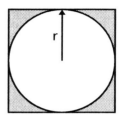

If the radius of the circle (r) is the same length as $\frac{1}{2}$ of the side of the square, then 2r × 2r = the area of the square.

Therefore, $(2r \times 2r) - \pi r^2 = 4r^2 - \pi r^2 = \boxed{r^2 (4 - \pi)}$

Write an expression in factored form for the area of the shaded region.

1
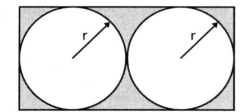

2
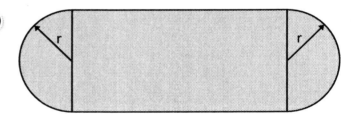

3
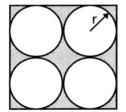

4
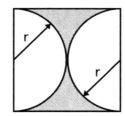

Factoring Perfect Square Trinomials

$x^2 - 8x + 16 = (x - 4)^2$
$4x^2 + 12x + 9 = (2x + 3)^2$

1 $(z - 1)^2$

4 $(3x - 2y)^2$

2 $(n + 5)^2$

5 $(12j - 10k)^2$

3 $(5k + 3)^2$

6 $(^-6m + n)^2$

Determine whether each trinomial is a perfect square. If it is, factor it. If it is not, then write "not a perfect square."

7 $x^2 + 12x + 6$

10 $a^2 + 2a + 1$

8 $4y^2 + 4y + 1$

11 $s^2 + 10s - 25$

9 $a^2 - 2ab + 4b$

12 $f^4 - 14f^2 + 49$

13 What do all of the last terms in a perfect square trinomial have in common?

_____

14 How can you tell what the sign will be (+ or −) in the factored form of the perfect square trinomial?

_____

Algebra © 2004 Creative Teaching Press

Factoring Pattern for $x^2 + bx + c$

$(x + 4)(x + 5) = x^2 + 9x + 20$

the sum of 4 & 5 ⟶   ⟵ the product of 4 & 5

1 $x^2 + 5x + 4$

2 $a^2 + 7a + 12$

3 $n^2 + 4n + 3$

4 $s^2 - 6s + 8$

5 $g^2 + 9g + 14$

6 $d^2 - 10d + 16$

7 $b^2 - 13b + 36$

8 $v^2 + 8v + 15$

9 $x^2 - 15x + 26$

10 $a^2 - 8a + 15$

11 $x^2 + 5x + 4$

12 $p^2 + 10p + 24$

13 $j^2 - 6j + 8$

14 $x^2 + 7x + 10$

15 $64 - 20b + b^2$

16 $3 - 4r + r^2$

17 $30 - 11q + q^2$

18 $26 - 15y + y^2$

Factoring Pattern for $x^2 + bx - c$

the sum of 4 & ⁻5 $\xrightarrow{\quad}$ $(x + 4)(x - 5) = x^2 - x - 20$ $\xleftarrow{\quad}$ the product of 4 & ⁻5

Remember, the coefficient of ⁻x is ⁻1!

1 $x^2 + 5x - 6$

2 $a^2 - 4a - 12$

3 $n^2 + 2n - 8$

4 $s^2 - 3s - 4$

5 $g^2 - 2g - 48$

6 $d^2 - 4d - 45$

7 $b^2 - 13b - 30$

8 $v^2 + 2v - 8$

9 $x^2 + 7x - 18$

10 $a^2 - 4a - 32$

11 $x^2 - 2x - 24$

12 $p^2 + 4p - 21$

13 $j^2 - 25j - 54$

14 $x^2 - 2x - 15$

15 $s^2 + 4s - 32$

16 $f^2 + 6f - 40$

17 $x^2 + xy - 56y^2$

18 $r^2 - 13rs - 48s^2$

Factoring Pattern for ax² + bx + c

$$(3x + 1)(x - 2) = 3x^2 - 5x - 2$$

$(3x)(x) = 3x^2$ ────────────────────┐

────── the product of 1 & ⁻2

$(3x)(⁻2) + (1)(x)$

$(⁻6x) + (x) = ⁻5x$ ──────────────

1 $5n^2 - 23n + 12$

2 $3a^2 + 7a + 2$

3 $2z^2 - 15w + 7$

4 $7b^2 + 61b + 40$

5 $5y^2 + 34y - 7$

6 $5y^2 - 23y + 12$

7 $5f^2 + 4f - 1$

8 $10c^2 - 59c - 6$

9 $5g^2 + 66g + 72$

10 $4a^2 - 3a + 9$

11 $3p^2 + 7p - 6$

12 $4c^2 + 4c - 3$

13 $2r^2 - 9r + 4$

14 $3q^2 + q - 4$

15 $10b^2 - 17b - 6$

16 $6s^2 - 19st + 3t^2$

17 $3e^2 + 11ef + 6f^2$

18 $6a^2 + 17ab + 10b^2$

Factoring by Grouping

$5(x + 2) - 2y(x + 2)$
$(5 - 2y)(x + 2)$

Hint: $(a - 2) = ^-(2 - a)$
$3(a - 2) - 2a(2 - a)$
$3(a - 2) - [^-2a(2 - a)]$
$(3 + 2a)(a - 2)$

1 $2(x - 3) + x(x - 3)$

6 $4(z + 3) + x(z + 3)$

2 $5(a + 3) + b(a + 3)$

7 $s(t - 3) + 5(3 - t)$

3 $3(r - 4) + s(r - 4)$

8 $m(n - 4) + 8(4 - n)$

4 $6(a + 7) + b(a + 7)$

9 $3a(6a - b) + 2(b - 6a)$

5 $e(f - g) - 4(f - g)$

10 $r(q - 1) - 7(1 - q)$

$2xy - 6xz + 3y - 9z = (2xy - 6xz) + (3y - 9z) = 2x(y - 3z) + 3(y - 3z)$
Answer: $(2x + 3)(y - 3z)$

11 $3xy + 12y + 4x + 16$

14 $7x^3 - 21x^2 - 10x + 30$

12 $3r + rs + 3t + st$

15 $24s^3 - 6s^2 - 20s + 5$

13 $ab + 5a + bc + 5c$

16 $a^3 - 2a^2 + 3a - 6$

Factoring Checklist

How do you know when you have factored completely?
1. Factor out the greatest monomial factor (GMF). Then . . . ✓
2. Decide whether it is a difference of squares. Then . . . ✓
3. Decide whether it is a perfect square trinomial. Then . . . ✓
4. Look for a pair of binomial factors. Then . . . ✓
5. Look for a way to group terms. Then . . . ✓
6. Check to determine if the polynomial is prime. Then . . . ✓
7. Multiply the factors to check your work. ✓

A. Answer the questions and use the steps to factor $x^2 - 121$.

❶ Is there a GMF? _____

❷ Is it a difference of squares? _____ What are the factors?

B. Answer the questions and use the steps to factor $7x^2 + 61x + 40$.

❶ Is there a GMF? _____

❷ Is it a difference of squares? _____

❸ Is it a perfect square trinomial? _____

❹ Is there a pair of binomial factors? _____ What are they?

C. Answer the questions and use the steps to factor the polynomial
$20n^3 + 24n^2 - 15n - 18$.

❶ Is there a GMF? _____

❷ Is it a difference of squares? _____

❸ Is it a perfect square trinomial? _____

❹ Is there a pair of binomial factors? _____

❺ Can you group the terms so that they can be factored? _____
What is the factored form of $20n^3 + 24n^2 - 15n - 18$?

Solving Equations by Factoring

$$(x - 6)(x + 3) = 0$$
$$x - 6 = 0$$
$$x = 6$$
$$x + 3 = 0$$
$$x = {}^-3$$
$$x = {}^-3 \text{ or } 6$$

1 $(x + 2)(x - 5) = 0$

2 $(n - 7)(n - 1) = 0$

3 $(a + 1)(a + 9) = 0$

4 $(d - 3)(d - 9) = 0$

5 $(s + 4)(s + 5) = 0$

6 $d^2 - 4d - 12 = 0$

7 $6b^2 - 3b - 3 = 0$

8 $3y(2y + 1)(2y + 5) = 0$

9 $3s^2 - 14s + 11 = 0$

10 $4n^2 - 15n = 4$

11 $6q^2 + 5q = 0$

12 $36a^2 = 1$

13 $n^2 = 20n - 100$

14 $f^2 = 16f$

15 $3s^2 + s = 2$

16 $4u^2 + 121 = 44u$

Algebra © 2004 Creative Teaching Press

Algebra Awareness (Factoring)

1 Find the greatest common factor of 18, 24, and 54.

   a) 12   b) 4   c) 6   d) 9

2 Find the greatest monomial factor of $24x^3y^2$ and $32x^4y$.

   a) $8x^3y$ b) $8xy$ c) $12xy^3$ d) $2x^3y$

Factor completely.

3 $a^4 - b^4$

   a) $a^3 - b^3$
   b) $(a - b)(a + b)$
   c) $a - b$
   d) $(a^2 + b^2)(a - b)(a + b)$

4 $a^2 - 8a + 16$

   a) $(a - 4)^2$
   b) $(a + 4)^2$
   c) $(a - 4)(a - 4)$
   d) $(a - 2)^2$

5 $n^2 - 6n + 8$

   a) $(n - 4)(n + 2)$
   b) $(n - 4)(n - 2)$
   c) $(n + 4)(n + 2)$
   d) $(n - 8)(y + 2)$

6 $y^2 - 2y - 8$

   a) $(y - 8)(y + 2)$
   b) $(y + 4)(y - 2)$
   c) $(y - 4)(y + 2)$
   d) $(y - 4)(y + 2)$

7 $6e^2 + 18ef + 12f^2$

   a) $(e - f)(e + f)$
   b) $(3e - f)(e + f)$
   c) $6(2e - 2f)$
   d) $6(e + 2f)(e + f)$

8 $x^2 - x^2y - y^2 + y^3$

   a) $(x + y)(x - y)$
   b) $(x - y)(x - y)$
   c) $(x + y)^2(x - y)$
   d) $(x + y)(x - y)(1 - y)$

Simplifying Algebraic Fractions

$$\frac{x^5}{x^2} = \frac{(x)(x)(x)(x)(x)}{(x)(x)} = x^3 \qquad \frac{3ab^2}{b^3} = \frac{3(a)(b)(b)}{(b)(b)(b)} = \frac{3a}{b}$$

1 $\dfrac{x^4}{x^3}$

2 $\dfrac{2a^5}{a^3}$

3 $\dfrac{5r^4}{r^2}$

4 $\dfrac{12d^4}{d}$

5 $\dfrac{24a^2b^2}{a^2b}$

6 $\dfrac{x^2y^3}{x^2y^2}$

7 $\dfrac{s^2}{s^2}$

8 $\dfrac{9u^6}{3u^3}$

9 $\dfrac{4a^2b}{8ab}$

10 $\dfrac{12x^2y^2}{x^4y^4}$

11 $\dfrac{48s^2t^5}{s^2t}$

12 $\dfrac{9m^2n^3}{3m^5n^4}$

13 $\dfrac{^-22e^4f^2}{11f^3}$

14 $\dfrac{^-15s^3}{^-r^3s^7}$

15 $\dfrac{^-19a^3b^4}{^-57a^2b^2}$

16 $\dfrac{18xy^6}{^-32x^5y^2}$

Dividing Polynomials by Monomials

$$\frac{6x^3 - 4xy^2 + 10xy}{2xy}$$

$$\frac{6x^3}{2xy} - \frac{4xy^2}{2xy} + \frac{10xy}{2xy}$$

$$\frac{(2)(3)xxx}{2xy} - \frac{(2)(2)xxy}{2xx} + \frac{(2)(5)xy}{2xx}$$

$$\frac{3x^2}{y} - 2y + 5$$

1 $\dfrac{6n + 9}{3}$

2 $\dfrac{24y - 12}{6}$

3 $\dfrac{12x - 6}{2}$

4 $\dfrac{4r - 2}{2}$

5 $\dfrac{9f + 3}{3}$

6 $\dfrac{24e + 6e}{3e}$

7 $\dfrac{7f - 21}{7}$

8 $\dfrac{10a + 5b}{5a}$

9 $\dfrac{33x^4 - 11x^3 - 44x^2}{11x}$

10 $\dfrac{8w^4 - 4w^3 - 6w^2}{^-2w^2}$

11 $\dfrac{7a^4b^3 - 7a^3b^2 - 7a^2b}{7}$

12 $\dfrac{8m^7n^5 - 8m^4n^2 - 2m^3n}{2m^2n}$

Name _____   Date _____

More Algebraic Fractions

$$\frac{x^2 - 2x + 4}{2 - x} = \frac{(x-2)(x-2)}{2-x} = \frac{(x-2)(x-2)}{^-(x-2)} = {}^-(x-2) = {}^-x + 2$$

1 $\dfrac{2x + 2y}{4}$

2 $\dfrac{5e - 10}{e - 2}$

3 $\dfrac{6b + 30}{b^2 - 25}$

4 $\dfrac{4r^2 - 8r}{4r^3}$

5 $\dfrac{7c - 7d}{c^2 - d^2}$

6 $\dfrac{y^2 + 8y + 16}{y^2 - 16}$

7 $\dfrac{25 - f^2}{f^2 + 12f + 35}$

8 $\dfrac{6s^3 - 24s^2}{s^2 + s - 20}$

9 $\dfrac{25x + 15y}{50x^2 + 30y^2}$

10 $\dfrac{2w^2 + w - 6}{2w + 4}$

11 $\dfrac{a^2 + 5a}{a^2 - 25}$

12 $\dfrac{4b^2 - 5b - 6}{8b^2 + 6b}$

13 $\dfrac{2n^2 - 9n + 4}{8n - 2n^2}$

14 $\dfrac{3x^2 - 4xy - 7y^2}{x^2 - y^2}$

Algebra © 2004 Creative Teaching Press

Multiplying Fractions

$$\frac{y^2 - 5y}{5} \cdot \frac{10y}{2y - 10} \;=\; \frac{10y^3 - 50y^2}{10y - 50} \;=\; \frac{10y^2(y - 5)}{10(y - 5)} \;=\; y^2$$

1 $\dfrac{5}{2} \cdot \dfrac{2}{3}$

2 $\dfrac{^-8}{15} \cdot \dfrac{3}{40}$

3 $\dfrac{a}{b} \cdot \dfrac{b}{c}$

4 $\dfrac{3}{xy} \cdot \dfrac{2x^2}{y}$

5 $\dfrac{n-1}{3} \cdot \dfrac{12}{n^2 - 1}$

6 $\dfrac{x+2}{x} \cdot \dfrac{x^2}{x^2 - 4}$

7 $\dfrac{a^2 - 1}{a} \cdot \dfrac{a^2}{a - 1}$

8 $\dfrac{a^2 - c^2}{a} \cdot \dfrac{4}{2c - 2a}$

9 $\dfrac{k^2 - 6k - 16}{k^2 + 4k - 21} \cdot \dfrac{k^2 - 8k + 15}{k^2 + 9k + 14}$

10 $\dfrac{x^2 + y^2}{x^2 + 2xy + y^2} \cdot \dfrac{3(x+y)}{6}$

11 $\dfrac{4x^2 - 25y^2}{(5y - 2x)^2} \cdot \dfrac{4}{4x + 10y}$

12 $\dfrac{4b^2 - 2ab}{b^2 - 4ab + 4a^2} \cdot \dfrac{(2a - b)^3}{2b - a}$

13 $\dfrac{5a + 5b}{a^2 - b^2} \cdot \dfrac{a^2 - ab}{(a + b)^2}$

14 $\dfrac{6a^2b - 4ab^2}{45a^2 - 20b^2} \cdot \dfrac{30a + 20b}{4a^2b^2}$

Area with Fractions

The area of a rectangle equals its length times its width $(A = LW)$.

The area of a triangle is half of the area of the same product $\left(A = \frac{1}{2}LW\right)$.

Find the area of each figure.

❶

$$\frac{3x}{7}$$

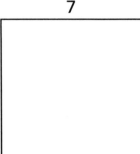

❷

$$\frac{x+6}{2}$$

$$\frac{x-6}{2}$$

Find the area of each shaded triangle.

❸

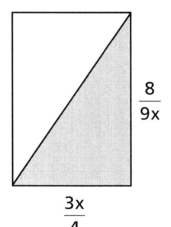

$$\frac{8}{9x}$$

$$\frac{3x}{4}$$

❹

$$\frac{4x}{3}$$

Dividing Fractions

$$\frac{x-y}{x+y} \div \frac{1}{(x+y)^2} \;=\; \frac{x-y}{x+y} \cdot \frac{(x+y)^2}{1} \;=\; \frac{x-y}{\cancel{x+y}} \cdot \frac{\cancel{(x+y)}(x+y)}{1} \;=\; (x-y)(x+y) = x^2 - y^2$$

1 $\dfrac{5}{8} \div \dfrac{25}{16}$

2 $\dfrac{3x}{y} \div \dfrac{x}{2}$

3 $\dfrac{3a^2 b}{4} \div \dfrac{6b}{a}$

4 $\dfrac{^-3n^2}{2m} \div \dfrac{2n^2}{3m}$

5 $\dfrac{e+2}{e^2-9} \div \dfrac{1}{e-3}$

6 $\dfrac{x-1}{2} \div \dfrac{x^2-1}{6}$

7 $\dfrac{n^2-4}{2n} \div (n+2)$

8 $\dfrac{2a+b}{a^2} \div \dfrac{a^2-b^2}{3a}$

9 $\dfrac{x^2-y^2}{x^2+2xy+y^2} \div \dfrac{x-y}{x+y}$

10 $\dfrac{x^2-y^2}{a+b} \div \dfrac{7x+7y}{7a+7b}$

11 $\dfrac{3}{a^2-9} \div \dfrac{3a-6}{a-3}$

12 $\dfrac{3-3r}{r^2+2r-3} \div (2r-2)$

13 $\dfrac{\dfrac{28x^2}{3x^2 y}}{\dfrac{14x^3}{9y^2}}$

Mix It Up!

Express answers in simplest form.

1 $\dfrac{x^2 - 2x}{x^2 - 3x - 4} \cdot \dfrac{x^2 - 25}{x^2 - 4x - 5} \div \dfrac{x^2 + 5x}{5x^2 + 10x + 5}$

2 $\dfrac{a^2 + 6a - 7}{6a^2 - 7a - 20} \cdot \dfrac{2a^2 + a - 15}{a^2 + 2a - 3} \div \dfrac{a^2 + 5a - 14}{3a^2 - 2a - 8}$

3 $\dfrac{2y + 2x - xy - x^2}{2 + y} \div \dfrac{y^2 - x^2}{2 + x} \cdot \dfrac{x - y}{x^2 - 4}$

4 $\dfrac{a^3 + 5a^2 - 9a - 45}{2a^2 - 9a + 9} \div \dfrac{a^2 + 10a + 25}{12 - 14a + 4a^2} \cdot \dfrac{a + 5}{a^2 - 4a + 4}$

5 $\dfrac{m^2 + 2mn + n^2 - 16}{16m^4 - 16n^4} \div \dfrac{m + n - 4}{4m^2 + 4n^2} \cdot \dfrac{m + n}{m + n + 4}$

6 $\dfrac{r^2 - 6rs + 9s^2 - 9}{r^4 - 81s^4} \cdot \dfrac{3r - 9s}{3r - 9s + 9} \div \dfrac{r - 3s - 3}{3r^2 + 27s^2}$

Homework Help

Hey look! Our friend Al Gebra is bringing up his grade. Yet, he still needs help. Would you help him again with his homework? Find and correct his 4 mistakes.

1 $\dfrac{9x + 9y}{3} = 3x + 3y$

2 $\dfrac{x^2 - 16}{x + 4} = x - 4$

3 $\dfrac{4s^2 - 5s - 6}{2s^2 + 4s} = \dfrac{4s + 3}{s - 2}$

4 $\dfrac{j}{k} \cdot \dfrac{k}{i} = \dfrac{j}{i}$

5 $\dfrac{x^2 - y^2}{x} \cdot \dfrac{4}{2x - 2y} = \dfrac{2x + 2y}{x}$

6 $\dfrac{28e + 4f}{e^2 - f^2} \cdot \dfrac{e^2 - ef}{7e^2 + f} = 1$

7 $\dfrac{6n^3 - 24n^2}{n^2 + n - 20} = \dfrac{6n^2}{n - 4}$

8 $\dfrac{3x}{y} \div \dfrac{x}{2} = \dfrac{6}{y}$

9 $\dfrac{^-8ab^2}{b} \div \dfrac{4a}{ab} = {}^-2ab^2$

10 $\dfrac{x^2 - y^2}{m + n} \div \dfrac{5x + 5y}{5m + 5n} = x - y$

11 $\dfrac{n^2 - 16}{m^4 - n^4} \div \dfrac{n - 4}{16m^2 + 16n^2} \cdot \dfrac{m + n}{n + 4} = 1$

Name _____ Date _____

Least Common Denominators

In order to add or subtract fractions, you must know the least common denominator (LCD). To find the LCD, factor completely. Find the product of the greatest power of each factor.

$$\frac{3}{6n-30}, \frac{8}{9n-45}$$

$$\frac{3}{(2)(3)(n-5)} \quad \frac{8}{(3)(3)(n-5)}$$

Therefore, the LCD is $2 \times 3^2 \times (n-5)$.

Find the least common denominator for each pair of fractions.

❶ $\dfrac{3}{2}, \dfrac{9}{10}$

❷ $\dfrac{1}{2}, \dfrac{1}{3}$

❸ $\dfrac{8}{x}, \dfrac{3}{xy}$

❹ $\dfrac{12}{x}, \dfrac{4}{xy^2}$

❺ $\dfrac{2n}{7a}, \dfrac{n}{2}$

❻ $\dfrac{1}{x}, \dfrac{1}{y}$

❼ $\dfrac{3}{x^2+1}, \dfrac{4}{x+1}$

❽ $\dfrac{3}{4a^2}, \dfrac{7}{12a}$

❾ $\dfrac{2a}{8a^2b}, \dfrac{n}{4ab^2}$

❿ $\dfrac{12}{8r^2st}, \dfrac{10}{2rs^2t}$

Adding and Subtracting Fractions with Like Denominators

$$\frac{2x+3y}{6y} - \frac{4x-y}{6y} \;=\; \frac{2x+3y-4x+y}{6y} \;=\; \frac{^-2x+4y}{6y} \;=\; \frac{\cancel{2}\left(^-x+2y\right)}{\cancel{2}(3y)} \;=\; \frac{^-x+2y}{3y}$$

1 $\dfrac{3x}{9} + \dfrac{2x}{9}$

6 $\dfrac{2f}{6} + \dfrac{f}{6}$

2 $\dfrac{6r}{4b} + \dfrac{8s}{4b}$

7 $\dfrac{s^2 - 7s}{(s-3)^2} + \dfrac{12}{(s-3)^2}$

3 $\dfrac{g}{g-h} - \dfrac{h}{g-h}$

8 $\dfrac{6y}{3} - \dfrac{y-7}{3}$

4 $\dfrac{x^2}{x+2} + \dfrac{4}{x+2}$

9 $\dfrac{x}{x-2} - \dfrac{2}{x-2}$

5 $\dfrac{3}{6a^2bc} + \dfrac{2}{6a^2bc}$

10 $\dfrac{8q-1}{5} - \dfrac{3q-6}{5}$

Name _____   Date _____

Fill in the Blanket

In order to add or subtract fractions, you must know the least common denominator (LCD). To find the LCD, factor completely. Find the product of the greatest power of each factor.

$$\frac{3}{2x} + \frac{2}{xy} \quad = \quad \frac{3(y)}{2x(y)} + \frac{2(2)}{xy(2)} \quad = \quad \frac{3y+4}{2xy}$$

Find the common denominators of the fractions below. Then add or subtract them together. Finally, match them with their correct answer(s).

$\frac{7}{3a} - \frac{4}{b}$		$\frac{3}{4} + \frac{2}{5}$		$\frac{4de+9f}{18def}$
	$\frac{3}{8}$		$1\frac{11}{20}$	
$\frac{23}{20}$		$\frac{12b-7f}{3ab}$		$1\frac{3}{20}$
	$\frac{12a-7b}{3ab}$		$\frac{5}{8}$	
$\frac{3}{6de} + \frac{2}{9f}$		$\frac{1}{5} + \frac{19}{20}$		$\frac{^-12a+7b}{3ab}$

68

Algebra © 2004 Creative Teaching Press

Adding and Subtracting Fractions with Unlike Denominators

$$\frac{x}{8}+\frac{3+12x}{18x} \;=\; \frac{x(9x)}{8(9x)}+\frac{(3+12x)(4)}{18x(4)} \;=\; \frac{9x^2}{72x}+\frac{12+48x}{72x} \;=\; \frac{9x^2+48x+12}{72x}$$

1 $\dfrac{2x}{y}+\dfrac{3}{2y}$

6 $\dfrac{n+2}{5n}+\dfrac{n-1}{n^2}$

2 $\dfrac{4a}{b}+\dfrac{7a}{2b}$

7 $\dfrac{k}{2(k-1)}-\dfrac{k+1}{k-1}$

3 $\dfrac{3z+2}{w-2}+\dfrac{5z}{2-w}$

8 $\dfrac{3(e-f)}{20}-\dfrac{5(e+f)}{12}$

4 $\dfrac{6}{3-2x}-\dfrac{4x}{2x-3}$

9 $\dfrac{a-2}{4a}+\dfrac{a+2}{a^2}$

5 $\dfrac{2r}{y}+\dfrac{1}{y^2}$

10 $\dfrac{5n-2}{12}-\dfrac{3(n-3)}{8}$

Perimeter and Fractions

Find the perimeter of each figure.

1

$$\frac{z+1}{4}$$

$$\frac{z}{3}$$

2

$$\frac{x}{2}$$

$$\frac{x}{3}$$

$$x$$

3

$$\frac{a}{6}$$ $$\frac{a}{6}$$

$$\frac{3a}{4}$$

$$\frac{3a}{4}$$ $$\frac{3a}{4}$$

$$\frac{a}{4}$$ $$\frac{a}{4}$$

Algebra © 2004 Creative Teaching Press

Mixed Expressions

$$5 + \frac{2x+1}{x+1} \;=\; \frac{5(x+1)}{1(x+1)} + \frac{2x+1}{x+1} \;=\; \frac{5x+5}{x+1} + \frac{2x+1}{x+1} \;=\; \frac{5(x+1)}{1(x+1)} + \frac{2x+1}{x+1} \;=\; \frac{5x+5}{x+1} + \frac{2x+1}{x+1} \;=\; \frac{7x+6}{x+1}$$

1 $2\dfrac{5}{8}$

6 $\dfrac{n+4}{n} + 9$

2 $9\dfrac{4}{7}$

7 $(f-2) + \dfrac{3}{f-2}$

3 $3 + \dfrac{5}{a}$

8 $2z + \dfrac{z-6}{3z-1}$

4 $3x - \dfrac{2}{x}$

9 $b^2 - \dfrac{b+1}{b-1}$

5 $\dfrac{h}{h-3} + 5$

10 $2x + 3y - \dfrac{2x^2 - y^2}{2x - 3y}$

Algebra Awareness (Fractions)

1 $\dfrac{4y^2 - 17y + 4}{8y - 2y^2}$

   a) $\dfrac{2y - 1}{2y}$ b) $-\dfrac{y - 4}{2y}$

   c) $-\dfrac{4y - 1}{2y}$ d) $\dfrac{y - 4}{2y}$

2 $\dfrac{p + 2}{p} \bullet \dfrac{p^2}{p^2 - 4}$

   a) $\dfrac{p}{p + 2}$ b) $\dfrac{p}{p - 2}$

   c) $\dfrac{p}{2 - p}$ d) $\dfrac{p}{2 - p}$

3 $\dfrac{2 + 2b}{6} \div \dfrac{1 + b}{9}$

   a) $\dfrac{1 + b}{2}$ b) $\dfrac{1 + b}{3}$

   c) 2   d) 3

4 $\dfrac{x^2 + y^2}{x^2 - y^2} \div \dfrac{1}{x + y}$

   a) $\dfrac{x^2 - y^2}{x - y}$ b) $\dfrac{x^2 + y^2}{x + y}$

   c) $\dfrac{x^2 - y^2}{x + y}$ d) $\dfrac{x^2 + y^2}{x - y}$

5 $\dfrac{n - 2}{4n} + \dfrac{n + 2}{n}$

   a) $\dfrac{5n - 6}{2n}$ b) $\dfrac{5n + 6}{4n}$

   c) $\dfrac{5n - 6}{4n}$ d) $\dfrac{5n + 6}{2n}$

Algebra © 2004 Creative Teaching Press

What's the Word with Age?

Solve the word problems.

1 Henry's father is three times as old as Henry. Six years ago, he was five times older. How old is Henry?

2 Larry is 3 times as old as Moe, and Curly is 16 years younger than Larry. One year ago, Larry's age was twice the sum of the ages of Moe and Curly. Find each man's present age.

3 Twenty-five years ago, Betty was five more than one-third as old as Steve was. Today, Steve is twenty-six less than two times the age of Betty. How old is Steve?

4 Beth is $8\frac{2}{3}$ older than Ben. The sum of their ages is 23 years. How old is Ben?

5 Terri would be one-half as old as Willie if Terri were four years older. Willie is eight less than three times as old as Terri. How old is Willie?

Name _____ Date _____

Ratios

The ratio of 9 to 2 can be written as $9 \div 2$, $\frac{9}{2}$, or $9{:}2$.

The ratio of 2 hours to 15 minutes = $\frac{2h}{15\,min} = \frac{120\,min}{15\,min} = \frac{8}{1}$

Express each ratio as a fraction in simplest form.

1 12 miles to 38 miles

2 5 inches to 2 feet

3 6 hours to 4 days

4 9 feet to 9 yards

5 12 ounces to 4 pounds

6 the ratio of wins to losses in a 24-game season with 16 losses and no ties

7 the ratio of girls to boys in a classroom with 36 students, including 14 girls

8 the ratio of wins to losses in a 42-game season with 16 losses and no ties

9 the ratio of senators in the United States Congress to states in the Union

What's the Word with Ratios?

Solve the word problems.

1 Find the two numbers in the ratio 9:16 whose sum is 50.

2 Find the two numbers in the ratio 3:4 whose sum is 115.

3 Together there are 180 players and coaches in the city basketball league. If the player to coach ratio is 9:1, how many players are there?

4 The perimeter of a rectangle is 50 meters. Find the dimensions of the rectangle if they are in the ratio 1:4.

5 The ratio of boys to girls in an algebra class is 3:2. There are 35 students in the entire class. How many boys and how many girls are there in the class?

6 At the championship high school baseball game there were 700 spectators. The ratio of men to women was 17:18. How many women watched the game?

7 Diem drives her car 18 mi/h faster than Kevin rides his bike. The ratio of the distance they can travel in 1 hour 30 minutes is 5:2. Find their rates of speed.

Name _____ Date _____

Proportions

An equation stating two ratios that are equal is a proportion, like 3:4 = 6:8 OR $\frac{3}{4} = \frac{6}{8}$

One way to check and see if the ratios are proportionate is to use the extremes (the numerator of the first ratio and the denominator of the second) and the means. For example:

$\frac{3}{4} \times \frac{6}{8}$ 24 = 24 $\frac{n}{15} \times \frac{4}{5}$ 5n = 60   n = 12

Solve the proportions.

1 $\frac{3}{4} = \frac{12}{n}$

2 $\frac{1}{m} = \frac{9}{36}$

3 $\frac{3p}{7} = \frac{4}{5}$

4 $\frac{6}{3} = \frac{4}{2r}$

5 $\frac{18q}{13} = \frac{36}{39}$

6 $\frac{27}{u} = \frac{9}{u-6}$

7 $\frac{6}{a+3} = \frac{5}{a+2}$

8 $\frac{5}{w} = \frac{4}{w-25}$

9 $^-4 = \frac{8a}{5}$

10 $2 = \frac{6+4n}{7}$

11 $\frac{18}{60} = \frac{b}{5b-1.5}$

12 $\frac{2y+1}{2y} = \frac{3}{2}$

What's the Word with Percents?

What percent of 180 is 36?

n% $\times$ 180 = 36   0.01n $\times$ 180 = 36 $\dfrac{1.8n}{1.8} = \dfrac{36}{1.8}$ n = 20%

1 What percent of 80 is 16?

2 What percent of 40 is 10?

3 What percent of 25 is 16?

4 What percent of 450 is 9?

5 18 is 60% of what number?

6 6.3 is 7% of what number?

7 63 is 150% of what number?

8 25% of what number is 270?

9 1.5% of what number is 6?

10 500% of what number is 90?

11 How much is the sales tax on a $16,500 car if the sales tax is 8%?

Equations with Fractional Coefficients

$$\frac{n}{3} + \frac{n}{5} = 8 \qquad (15)\frac{n}{3} + (15)\frac{n}{5} = 8(15) \qquad 5n + 3n = 120 \qquad n = 15$$

1 $\quad \dfrac{x}{3} + \dfrac{x}{2} = 5$

2 $\quad \dfrac{e}{3} - \dfrac{e}{6} = 1$

3 $\quad \dfrac{3y}{4} + \dfrac{y}{2} = 5$

4 $\quad \dfrac{k}{3} + \dfrac{2k}{5} = \dfrac{7}{15}$

5 $\quad \dfrac{2z}{3} - \dfrac{z}{2} = 2$

6 $\quad \dfrac{5p}{12} - \dfrac{2p}{9} = 7$

7 $\quad \dfrac{2b}{5} + \dfrac{b}{3} = 0$

8 $\quad \dfrac{3-y}{2} = \dfrac{3}{4}$

9 $\quad \dfrac{s+3}{6} + \dfrac{s}{4} = 3$

10 $\quad \dfrac{3}{2}n + \dfrac{2}{6}n = 0$

11 $\quad \dfrac{3}{8}v - \dfrac{1}{4}v = 2$

12 $\quad \dfrac{6}{7}d + \dfrac{1}{2}d + 1 = 0$

Fractional Equations

$$\frac{3}{10w} - \frac{3}{5} = \frac{2}{5w} \qquad (10w)\frac{3}{10w} - (10w)\frac{3}{5} = \frac{2}{5w}(10w) \qquad 3 - 6w = 4 \qquad ^-6w = 1 \qquad w = -\frac{1}{6}$$

Solve. If the problem has no solution, write **no solution**.

1 $\dfrac{1}{5} + \dfrac{1}{x} = \dfrac{6}{5}$

7 $\dfrac{x}{x-2} = \dfrac{4}{3}$

2 $\dfrac{3}{m} - \dfrac{1}{4} = \dfrac{3}{4}$

8 $\dfrac{n}{n-2} = \dfrac{4}{5}$

3 $\dfrac{1}{r} + \dfrac{1}{3} = \dfrac{1}{2}$

9 $\dfrac{1+b}{b} = \dfrac{3}{b}$

4 $\dfrac{1}{3} + \dfrac{1}{g} = \dfrac{8}{15}$

10 $\dfrac{7}{z} - \dfrac{4z}{2z-3} = {}^-2$

5 $\dfrac{1}{2} - \dfrac{1}{s} = \dfrac{3}{10}$

11 $\dfrac{2x}{x+3} + 3 = \dfrac{x}{x+3}$

6 $\dfrac{5}{6y} + 3 = \dfrac{1}{2y}$

12 $\dfrac{1}{a} - \dfrac{2a}{a+1} = 0$

Mixture Problems

Solve the word problems.

1 How much water should be added to one liter of pure ammonia to make a mixture of 50% ammonia?

2 How many gallons of water must be added to 50 gallons of a 30% acid solution in order to produce a 20% acid solution?

3 Jessica's piggy bank has 100 coins, including pennies, nickels, dimes, and quarters. Together they are worth $8.36. There are nine more dimes than pennies and five times as many nickels as pennies. How many of each kind of coin does Jessica have in her piggy bank?

4 If Sally works overtime, she is paid an extra 50% per hour. After working her usual 40 hour week, she worked an additional 4 hours overtime. If she made $552 last week, what is her hourly wage?

5 Donna has 40 ounces of an alloy containing 65% iron. How many ounces of a second alloy that is 42% iron should be mixed with the first alloy to get a new alloy that is 50% iron?

Algebra © 2004 Creative Teaching Press

Homework Help

Oh no! Our friend Al Gebra is slipping again. He needs our help more than ever. Find and correct his 3 mistakes.

❶ What is the ratio of wins to losses in a 48-game season with 16 wins and no ties?

1:2

❷ The perimeter of a rectangle is 56 meters. Find the dimensions of the rectangle if they are in the ratio 1:3.

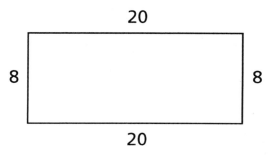

20

8   8

20

Solve the proportions.

❸ $\dfrac{24}{r} = \dfrac{8}{r-2}$

$r = 2$

❹ $\dfrac{3}{8}b - \dfrac{1}{4}b = 3$

$b = 24$

❺ $\dfrac{1}{n} + \dfrac{1}{3} = \dfrac{1}{12}$

$n = 4$

Algebra © 2004 Creative Teaching Press

Equations in Two Variables

(4, 0) is an ordered pair. $2x + 3y = 8$ is an equation with two variables. In the ordered pair (4, 0), 4 is the value of x and 0 is the value of y. Also, (4,0) is a solution of the equation because when you substitute their values, the equation $2(4) + 3(0) = 8$ is a true statement.

State whether each ordered pair is a solution of the equation.

1 $x - y = 5$
(9, 4) (7, 3)

5 $6x - 12y = 15$
(2, 1) (0.5, ⁻1)

2 $y = 4x + 2$
(1, 2) (1, 6)

6 $2xy = 8$
(2, 2) (3, 3)

3 $x + y = 6$
(9, 4) (7, 3)

7 $x^2 + y^2 = 25$
(3, 4) (⁻3, ⁻4)

4 $x - y = 5$
(4, 3) (1, 5)

8 $3x + 5y = 8$
(2, 1) (0.5, ⁻1)

Complete each ordered pair to form a solution of the equation.

9 $y = x - 3$ (8, ?), (3, ?)

13 $2x + y = 12$ (1, ?), (⁻1, ?)

10 $y = 0.5x - 2$ (4, ?), (2, ?)

14 $x - y = 4$ (⁻1, ?), (0, ?)

11 $y = \dfrac{3}{x+1}$ (5, ?), (⁻4, ?)

15 $2xy + 3y = 6$ (3, ?), (⁻3, ?)

12 $x + y = 8$ (4, ?), (⁻3, ?)

16 $2x - 3y = 12$ (0, ?), (1, ?)

Graphing Coordinates

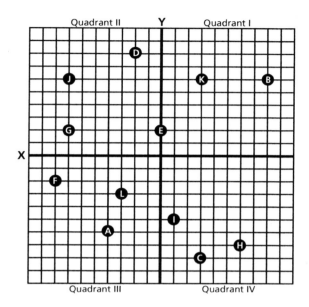

I. Give the coordinates of each point.

1 A _____   **3** E _____   **5** I _____

2 C _____   **4** G _____   **6** K _____

II. Name the point which is the graph of each ordered pair.

7 (6, ⁻7) _____   **9** (⁻7,6) _____   **11** (8, 6) _____

8 (⁻2, 8) _____   **10** (⁻8, ⁻2)_____   **12** (⁻3, ⁻3) _____

13 In what direction do the quadrants advance? Clockwise or counter-clockwise?

_____

III. Name the quadrant or axis of each ordered pair.

14 (0, ⁻2) _____   **16** (⁻6, ⁻4)_____   **18** (7, 0) _____

15 (⁻2, 5) _____   **17** (1, 1) _____

Slope of a Line

$$\text{slope} = \frac{\text{rise}}{\text{run}} = \frac{\text{vertical change}}{\text{horizontal change}} = \frac{\text{difference between y coordinates}}{\text{difference between x coordinates}}$$

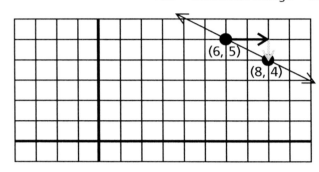

(6, 5)

(8, 4)

rise = ⁻1

run = 2

The slope of (8, 4) and (6, 5) = $\frac{4-5}{8-6}$ = ⁻$\frac{1}{2}$

❶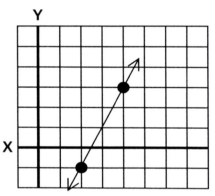

❷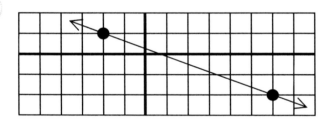

❹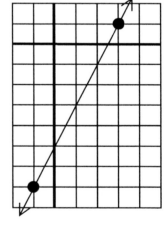

❸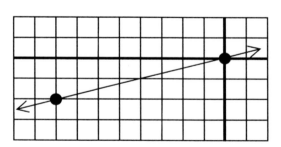

Find the slope of the line which passes through the given points.

❺ (⁻4, 2), (⁻6, 5)

❼ (4, 8), (1, 3)

❻ (0, ⁻3), (3, ⁻1)

❽ (⁻8, ⁻7), (⁻6, ⁻2)

Algebra © 2004 Creative Teaching Press

Slope Sleuth

(3, 2) and (6, -1) $\longrightarrow$ $\dfrac{2}{3} - \dfrac{(^-1)}{6} = \dfrac{3}{^-3} = ^-1$ $\longrightarrow$
$$y = mx + b$$
$$y = (^-1)x + b$$
$$y = (^-1)x + b$$

Choose either (3, 2) or (6, -1). $\quad 2 = (^-1)\,3 + b$
$$b = 5$$
$$y = {}^-x + 5$$

Change it to standard form. $\quad y + x = {}^-x + x + 5$
$$x + y = 5$$

The coordinate plane system is sometimes called by another name in honor of a 17th century mathematician and thinker. Follow the clues to find the name.

Find the slopes. Then write an equation in standard form of the line passing through the given points. Match the ordered pairs in Column A with the standard form in Column B. Write the letter in the correct space below.

Column A

❶ (0, 3), (2, -1)

❷ (5, 2), (7, 0)

❸ (0, 7), (1, 9)

❹ (-1, -2), (0, 3)

❺ (8, 1), (1, 8)

❻ (1, 4), (2, 5)

❼ (-2, 0), (2, -3)

❽ (0, -1), (1, 4)

Column B

Ⓣ $^-x + y = 3$

Ⓔ $^-5x + y = ^-1$

Ⓒ $x + y = 7$

Ⓐ $x + y = 9$

Ⓡ $2x + y = 3$

Ⓘ $3x + 4y = ^-6$

Ⓢ $^-2x + y = 7$

Ⓝ $^-5x + y = 3$

___ ___ ___ ___ ___ ___ ___ ___ ___
2   5   1   6   8   3   7   5   4

The Slope-Intercept Form

slope ————————→ $y = mx + b$ ←———————— y intercept

$$^-2x - y = 7$$
$$^-2x + 2x - y = 7 + 2x$$
$$^-y = 2x + 7$$
$$^-y = ^-2x - 7$$

slope = $^-2$ y intercept = $^-7$

Solve for y. Then state the *m* and the *b*.

1 $5x + y = {}^-3$

2 $3x + y = 8$

3 $^-x = y - 5$

4 $3y = 4x + 7$

5 $2y - 6x = 2$

6 $6y = 3x + 12$

Solve for y. Use the *m* and the *b* to graph the line.

7 $4x - 3y = 9$
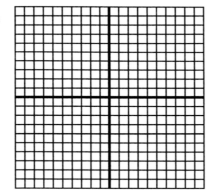

8 $x + 5y = 5$
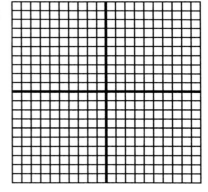

9 $3x - y = 6$
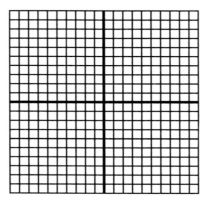

10 $y + {}^-2x = 3$
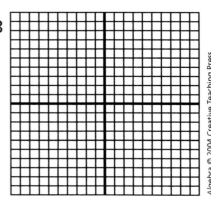

More Practice with the Slope-Intercept Form

Solve for y. Then graph each equation using the slope and the y-intercept.

1 2x + y = 5

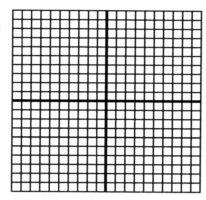

2 x − y = 4

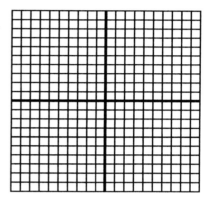

3 x + 3y = 6

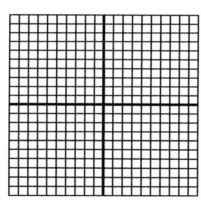

4 x = 4

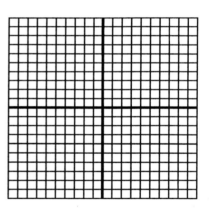

5 2x + y = 0

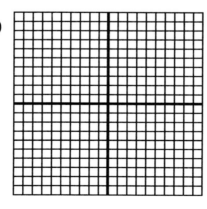

6 x − y = 6

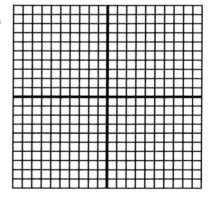

Algebra Awareness (Graphing)

Which pair is a solution of the equation?

1 $y = 3x + 5$

  a) (1, 2)   b) (5, 6)

  c) (1, ⁻2)   d) (⁻1, 2)

2 $y = {}^-2x - 4$

  a) (2, 0)   b) (⁻2, 0)

  c) (0, 2)   d) (0, ⁻2)

3 The quadrants advance in what direction?

  a) left   b) right

  c) clockwise   d) counter-
                      clockwise

4 What is the slope of (3, ⁻5) and (⁻2, 0)?

  a) ⁻1   b) 0 slope

  c) 1   d) no slope

5 What is the slope of $y = 3x - 2$?

  a) ⁻3   b) ⁻2

  c) 2   d) 3

6 What is the y-intercept of $y = {}^-5x - 4$?

  a) 4   b) ⁻5

  c) 5   d) ⁻4

7 Rewrite $y = {}^-2x + 5$ in standard form.

  a) ⁻2x + y = 5   b) 2x − y = 5

  c) 2x + y = ⁻5   d) 2x + y = 5

8 Rewrite $5x - 3y = 4$ in slope-intercept form.

  a) $y = -\dfrac{3}{5}x + \dfrac{4}{3}$ b) $y = -\dfrac{5}{3}x - \dfrac{4}{3}$

  c) $y = \dfrac{3}{5}x + \dfrac{4}{3}$ d) $y = \dfrac{5}{3}x - \dfrac{4}{3}$

Translating Word Problems with Multiple Unknowns

The sum of two numbers is 15. 3 times one of the numbers is 11 less than 5 times the other. What are the numbers?

n = one number   (15 – n) = the other number
   3n = 5n – 11
   3n = 5(15 – n) – 11   n = 8
   3n = 75 –5n – 11   15 – n
3n + 5n = 64 – 5n + 5n   15 – 8 = 7
   8n/8 = 64/8
    n = 8   the numbers are 7 and 8

1 4 times a number increased by 25 is 13 less than 6 times the number. What is the number?

2 3 times a number decreased by 8 is the same as twice the number increased by 15. What is the number?

3 One number is 28 more than 3 times another number. If each number were multiplied by 4, their difference would be 232. What are the numbers?

4 A number is 3 less than 4 times another number. Their sum is 102. What are the numbers?

5 If the larger of two numbers were decreased by 349, then the two numbers would be the same. The sum of the two numbers is 735. What are the numbers?

6 The larger of two numbers is 6 more than 6 times the smaller number. The larger number is also 122 more than 2 times the smaller number. What are the numbers?

Systems of Linear Equations: The Graphing Method

Find a solution to the system of equations by graphing.

$$x + 2y = 5 \longrightarrow y = -\frac{1}{2}x + \frac{5}{2}$$

$$x - y = 2 \longrightarrow y = x - 2$$

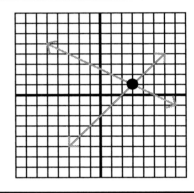

Solution: (3, 1)

❶ $x + y = 6$
$x - y = 0$

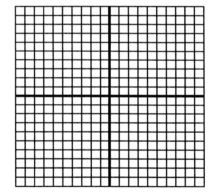

❷ $x - y = 4$
$x + y = 0$

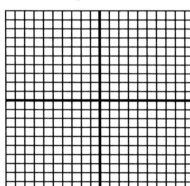

❸ $2x - y = 8$
$x + y = 1$

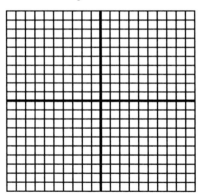

❹ $x - y = 3$
$2x + y = 3$

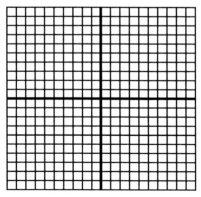

❺ $5x + 2y = 10$
$5x - 2y = 10$

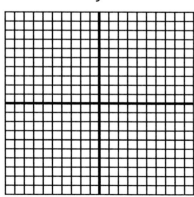

❻ $x + y = {}^-8$
$3x + y = {}^-6$

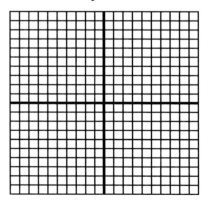

Systems of Equations:
The Substitution Method

$$y = 2x$$
$$5x - y = 30$$
$$5x - 2x = 30$$
$$3x = 30$$
$$x = 10$$

$$y = 2x$$
$$y = 2 \, (10)$$
$$y = 20$$

solution = (10, 20)

1 $y = 6x$
$x + y = 28$

8 $a = {}^-b + 1$
$a = {}^-5b + 5$

2 $n = 3m$
$n - m = 12$

9 $y = {}^-x + 9$
$y - 4x = {}^-31$

3 $a = 4b$
$3a - 4b = 20$

10 $2s + 5r = {}^-9$
$s = {}^-4r - 12$

4 $5y + x = 40$
$y = 3x + 40$

11 $5x = {}^-y - 15$
$x = 3y + 29$

5 $y = {}^-3x + 5$
$2y - 5x = 10$

12 $a = {}^-2b + 8$
${}^-a - 3b = {}^-10$

6 ${}^-6 = x - 2y$
$y = {}^-x + 6$

13 $h = {}^-f + 5$
$h + 4f = 5$

7 $y = 4x - 10$
${}^-5x = {}^-50 + 5y$

14 $x = {}^-3y + 35$
$4x - 3y = {}^-10$

Systems of Equations: The + or − Method

$$x + 4y = ^-2$$
$$\underline{+\ \ ^-x - 2y = 0}$$
$$2y = ^-2$$
$$y = ^-1$$
$$x + 4y = ^-2$$
$$x + 4(^-1) = ^-2$$
$$x + ^-4 = ^-2$$
$$x = 2$$
$$\text{solution} = (2,\ ^-1)$$

$$5x - 2y = ^-2$$
$$\underline{-\ \ 5x + 3y = 3}$$
$$^-5y = ^-5$$
$$y = 1$$
$$5x + 3y = 3$$
$$5x + 3(1) = 3$$
$$5x + 3 = 3$$
$$x = 0$$
$$\text{solution} = (0,\ 1)$$

1 $x + y = 7$
$x - y = 3$

2 $r + s = 5$
$r - s = 7$

3 $4x + 4y = 4$
$^-4x - 2y = ^-10$

4 $3x + 4y = 25$
$3x - 3y = 18$

5 $^-4x + y = ^-16$
$4x + y = 16$

6 $2x - 2y = 0$
$4x - 2y = 6$

7 $3x - y = 32$
$4x + y = 24$

8 $a = ^-b + 1$
$a = ^-5b + 5$

9 $4x - 5y = ^-15$
$5x - 5y = ^-15$

10 $x + 4y = 1$
$5x - 4y = 5$

11 $x + 5y = 5$
$4x + 5y = 5$

12 $x + 4y = 2$
$^-x - 5y = ^-1$

13 $2x - 5y = ^-1$
$^-2x + y = ^-3$

14 $x - 3y = ^-20$
$4x - 3y = 1$

Systems of Equations:
Multiplication with the + or – Method

$$2x + y = 8$$
$$\underline{3x - 2y = 5}$$
$$\overline{(2)2x + (2)y = 8(2)}$$
$$\underline{3x - 2y = 5}$$
$$\overline{4x + 2y = 16}$$
$$\underline{+ \; 3x - 2y = 5}$$
$$7x = 21$$
$$x = 3$$

$$2x + y = 8$$
$$2(3) + y = 8$$
$$6 + y = 8$$
$$y = 2$$

solution = (3, 2)

1 $3x - 2y = 4$
    $2x + y = 5$

2 $3r + 5s = 3$
    $r + 2s = 13$

3 $3x - 5y = 4$
    $x - 10y = {}^-7$

4 $2x - y = 0$
    $^-2x - 3y = 0$

5 $^-2x + y = 9$
    $x + 2y = {}^-22$

6 $x + 2y = 4$
    $^-3x - 8y = {}^-24$

7 $x + y = {}^-2$
    $^-2x + 3y = 24$

8 $4x + 4y = {}^-16$
    $2x + 8y = 10$

9 $^-x - 5y = {}^-13$
    $2x + 2y = {}^-6$

10 $2x - 5y = {}^-16$
    $x + 20y = {}^-8$

11 $2x - 3y = {}^-2$
    $5x + 9y = 28$

12 $3x + 3y = 12$
    $6x + 2y = 12$

13 $^-3x - 5y = {}^-21$
    $^-9x + 2y = {}^-63$

14 $^-3x + 3y = 12$
    $^-6x - y = {}^-39$

Puzzle Problems

Jim is 3 years older than Darlene. In 7 years Jim will be twice as old as Darlene was 5 years ago. How old are they now?

$$J = D + 3$$
$$J + 7 = 2(D - 5)$$
$$J + 7 = 2D - 10$$
$$(D + 3) + 7 = 2D - 10$$
$$D + 10 = 2D - 10$$
$$D + 20 = 2D$$
$$20 = D$$

$$J = D + 3$$
$$J = 20 + 3$$
$$J = 23$$

Darlene is 20.
Jim is 23.

1 Denise is 6 years older than her sister. 8 years ago she was 4 times as old as her sister. How old is each girl now?

2 6 years ago Cathy was twice as old as Kitty. In 6 years Kitty will be as old as Cathy is now. Find their ages now.

3 The sum of the digits of a number is 8. If the digits are reversed, the number is decreased by 54. What is the original number?

4 The sum of the digits of a number is 15. If the digits are reversed, the number is increased by 27. Find the number.

5 The denominator of a fraction is 7 more than the numerator. If 5 is added to each, the result is $\frac{1}{2}$. Find the original fraction.

6 The denominator of a fraction is 1 more than the numerator. If the numerator is decreased by 1, the resulting fraction is $\frac{3}{4}$. Find the original fraction.

Name _____ Date _____

Homework Help

Our friend Al Gebra is doing better. But he still needs some help. Find and correct his 2 mistakes.

1 The point at which two linear equations intersect is called the _____ solution _____ of the system of equations.

2 Solve using the substitution method.

$$x = {}^-2y + 8$$
$${}^-x - 3y = {}^-10$$

(2, 4)

3 Solve using the + method.

$$x - y = {}^-5$$
$$+ \quad {}^-x - 3y = {}^-5$$

(0, 5)

4 Solve using the − method.

$$3x + 5y = {}^-10$$
$$- \quad 3x + 2y = 14$$

(10, ⁻8)

5 Solve using the multiplication with + or − method.

$$2x + 3y = 7$$
$${}^-4x + 5y = {}^-3$$

(2, 1)

6 Hank is 5 more than 4 times Bethany's age. 5 years ago Hank was 9 times Bethany's age. How old are they now?

Hank = 38   Bethany = 7

Name _____ Date _____

Inequalities

The real numbers greater than -3

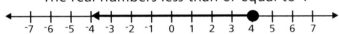

The real numbers less than or equal to 4

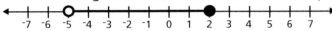

The real numbers greater than -5 and less than or equal to 2

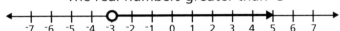

I. Graph the inequalities.

1 The real numbers less than -1

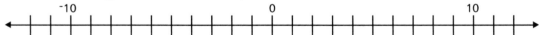

2 The real numbers greater than or equal to -2

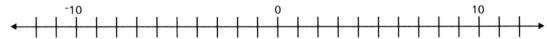

3 The real numbers greater than -6 and less than or equal to 3

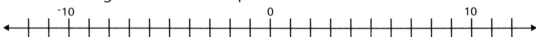

4 The real numbers greater than or equal to -1 and less than 1

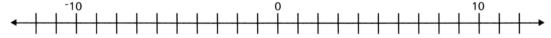

II. Classify each statement as true or false.

5 5 ≥ -3 _____

6 -2 < -4 _____

7 12 ≥ 12 _____

8 -6 < -4 _____

9 4 ≥ 1 _____

10 -8 ≥ -9 _____

96

Algebra © 2004 Creative Teaching Press

Solving Inequalities

$$7 \geq {}^-2n - 7$$
$$7 + 7 \geq {}^-2n - 7 + 7$$
$$14 \geq {}^-2n$$
$$\frac{14}{{}^-2} \geq \frac{{}^-2n}{{}^-2}$$
$${}^-7 \leq n$$

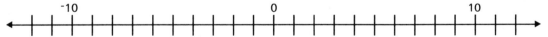

I. Solve the inequalities.

1 $n - 2 \geq 6$

2 $y + 3 > 7$

3 $2 - p < 0$

4 $z + 4 \leq 12$

5 $\dfrac{x}{5} < {}^-4$

6 $6 < 3b - 3$

7 $^-4k + 5 \geq 13$

8 $4 - 2m > 10$

II. Solve and graph the inequalities.

9 $4 > d + 4$

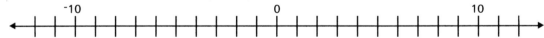

10 $3 \leq {}^-2w - 5$

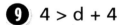

11 $\dfrac{x}{^-2} \geq 3$

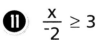

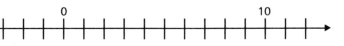

Algebra © 2004 Creative Teaching Press

Combined Inequalities

$$^-2 \leq 3 + n < 4$$
$$^-2 - 3 \leq 3 - 3 + n < 4 - 3$$
$$^-5 \leq n < 1$$

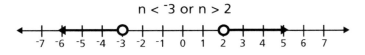

$$n < {}^-3 \text{ or } n > 2$$

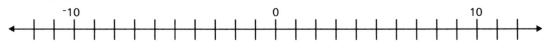

Solve and graph the inequalities.

1 $x > 2 \text{ or } x \leq {}^-3$

2 $^-5 < y < 2$

3 $^-2 < {}^-2a \leq 4$

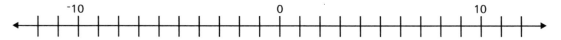

4 $n + 5 \leq {}^-2 \text{ or } n + 5 \geq 2$

5 $^-2 \leq 2a + 4 < 8$

Algebra © 2004 Creative Teaching Press

Absolute Value and Open Sentences

$|n| = 2$ therefore . . . $n = 2$ or $n = ^-2$
$|n| > 2$ therefore . . . $n < ^-2n$ or $n > 2$
$|n| < 2$ therefore . . . $^-2 < n < 2$

$$|a + 1| \leq 4$$
$$^-4 \leq a + 1 \leq 4$$
$$^-4 - 1 \leq a + 1 - 1 \leq 4 - 1$$
$$^-5 \leq a \leq 3$$

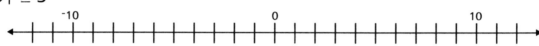

Solve and graph the inequalities.

1 $|45| \geq 3$

2 $|n - 4| < 3$

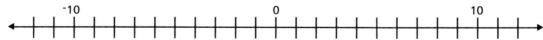

3 $|b + 4| \geq 5$

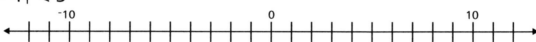

4 $|2x - 3| > 3$

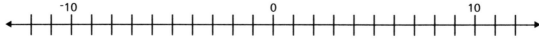

5 $|2n - 3| = n + 4$

Graphing Linear Inequalities

$$x - y > 2 \longrightarrow y < x - 2$$

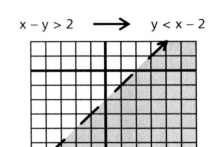

$$x + 2y \leq 5 \longrightarrow y \leq -\frac{1}{2}x + \frac{5}{2}$$

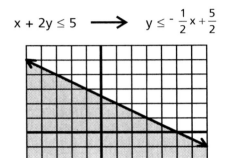

Solve and graph the inequalities.

1 $y \leq x + 3$

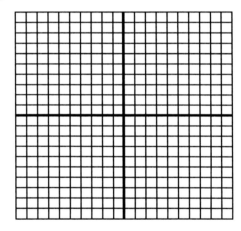

2 $y < {}^-2x - 2$

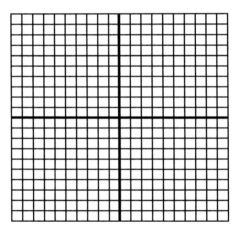

3 $2x + y \geq 1$

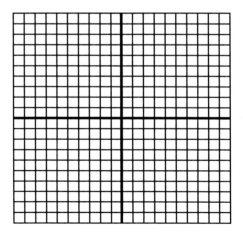

4 ${}^-3x - y > 2$

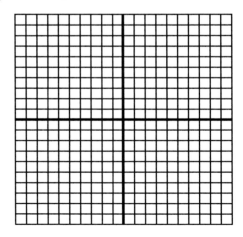

Algebra © 2004 Creative Teaching Press

Name _____ Date _____

Systems of Linear Inequalities

The solution set of the system

$$y > x - 2$$

and

$$y \leq -\frac{1}{2}x + \frac{5}{2}$$

is a graph containing both linear inequalities.

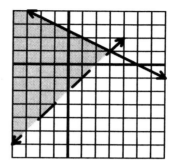

Solve and graph the inequalities.

1 $y \leq 3$ and $y < x - 2$

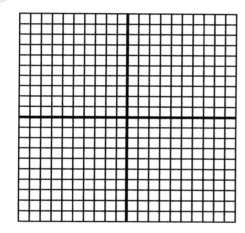

2 $y < {}^-x - 2$ and $y > {}^-3x + 2$

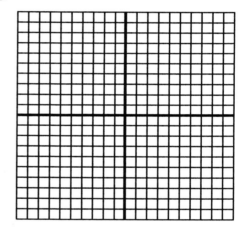

3 $4x + y \geq 1$ and $2x - y > 3$

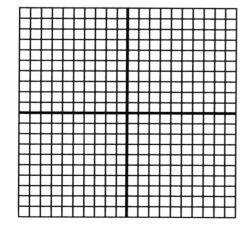

4 ${}^-3x - y > 2$ and ${}^-4x + 2y \leq 8$

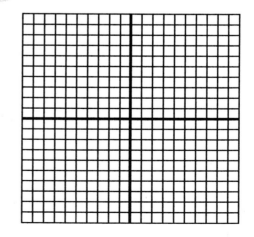

Name _____ Date _____

Algebra Awareness (Inequalities)

1 The figure below represents the real numbers greater than ⁻1 and _____.

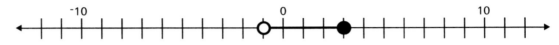

    a) less than 3   b) more than 3   c) less than ⁻3   d) less than or equal to 3

Solve the inequalities.

2 n + 2 ≥ 3

    a) n > 1   b) 1 ≥ n

    c) n ≥ 1   d) n ≤ 1

3 |2x + 1| > 9

    a) 4 < x < ⁻5   b) ⁻4 > x < ⁻5

    c) 4 < x < 5   d) 4 > x > ⁻5

4 Which inequality is represented in the graph?

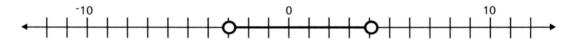

    a) 3 < y < 4   b) ⁻3 < y < 4   c) ⁻3 > y > 4   d) ⁻3 < y > 4

5 Which system of inequalities is represented below?

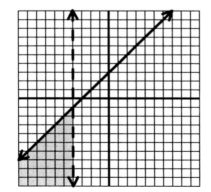

    a) y < ⁻4   b) x < 4
       y ≥ x + 3   y < x + 3

    c) x > ⁻4   d) x < ⁻4
       y ≥ x + 3   y ≤ x + 3

Algebra © 2004 Creative Teaching Press

Radicals

$$\sqrt{\frac{144}{36}} \quad = \quad \sqrt{\frac{(12)(12)}{(6)(6)}} \quad = \quad \frac{12}{6} \quad = \quad 2$$

1 $\sqrt{16}$

2 $\sqrt{49}$

3 $\left(\sqrt{4}\right)^2$

4 $\left(\sqrt{121}\right)^2$

5 $^-\left(\sqrt{43}\right)^2$

6 $\sqrt{\dfrac{1}{64}}$

7 $^-\sqrt{196}$

8 $\sqrt{\dfrac{36}{81}}$

9 $\pm\sqrt{49}$

10 $\pm\sqrt{169}$

11 $\pm\sqrt{1225}$

12 $\pm\sqrt{\dfrac{4}{36}}$

13 $\pm\sqrt{\dfrac{1}{256}}$

14 $^-\sqrt{\dfrac{81}{9}}$

15 $^-\sqrt{\dfrac{484}{100}}$

16 $^-\sqrt{\dfrac{324}{729}}$

Name _____ Date _____

Is That Rational?

A **rational number** is any integer, fraction, mixed number, or decimal, whether positive or negative.

Irrational numbers are real numbers that cannot be expressed in the form $\frac{a}{b}$ when a and b represent integers. Such an example is $\sqrt{2}$ because there is no whole number, fraction, or decimal whose square is 2.

Shade in the boxes that contain **rational numbers**.

3	$\sqrt{64}$	0.79	$\sqrt{25}$	0.79	$\sqrt{1.44}$	2
$\sqrt{\dfrac{28}{63}}$	$\left(\sqrt{2}\right)^4$	7.3	$\sqrt{16}$	2	$\sqrt{9}$	2
0.123	$\sqrt{\dfrac{243}{147}}$	$6.7\overline{8}$	5	19.1	$\sqrt{50}$	$7 - \sqrt{11}$
$\sqrt{3}$	$\sqrt{225}$	4	$\sqrt{36}$	$5 + \sqrt{7}$	2.9	$\sqrt{7} - \sqrt{7}$
11	7π	$\sqrt{100}$	1.23…	0.9	13	$\sqrt{4}$
$\sqrt{1.69}$	23	$\sqrt{2}$	$\sqrt{\dfrac{1}{36}}$	2	$2.\overline{91}$	$\sqrt{\dfrac{4}{25}}$
5.23	$\sqrt{7 \bullet 7}$	3.4	2.1	$\sqrt{3} - \sqrt{3}$	17	$\sqrt{9 \bullet 9}$

Algebra © 2004 Creative Teaching Press

Square Roots with Variables

$$\sqrt{121b^2} \quad = \quad \sqrt{(11)(11)bb} \quad = \quad 11\,|b|$$

1 $\sqrt{25x^2}$

2 $\sqrt{121b^2}$

3 $\sqrt{81n^2}$

4 $\sqrt{36z^4}$

5 $\sqrt{64r^2s^2}$

6 $\sqrt{100m^4n^4}$

7 $\sqrt{9c^4}$

8 $\sqrt{80b^2c^2}$

9 $\sqrt{144h^2}$

10 $\sqrt{49r^6s^6}$

11 $\sqrt{75q^3}$

12 $\pm\sqrt{56a^6b^4}$

13 $\sqrt{\dfrac{f^2}{49}}$

14 $\sqrt{\dfrac{e^4f^4}{169}}$

15 $\sqrt{\dfrac{9}{z^4}}$

16 $\pm\sqrt{54m^2n^3}$

17 $\sqrt{80v^6}$

18 $\pm\sqrt{\dfrac{100y^{10}}{64}}$

Multiplying Radicals

$$\sqrt{4} \times \sqrt{9} \ = \ \sqrt{36} \ = \ 6$$

$$2\sqrt{8} \times 3\sqrt{2} \ = \ (3 \times 2)(\sqrt{8} \times \sqrt{2}) \ = \ 6\sqrt{16} \ = \ 6 \times 4 \ = \ 24$$

1 $\sqrt{5} \cdot \sqrt{2}$

2 $\sqrt{3} \cdot \sqrt{6}$

3 $\sqrt{1} \cdot \sqrt{4}$

4 $\sqrt{9} \cdot \sqrt{4}$

5 $\sqrt{8} \cdot \sqrt{8}$

6 $\sqrt{18} \cdot \sqrt{8}$

7 $\sqrt{50} \cdot \sqrt{2}$

8 $\sqrt{3} \cdot \sqrt{12}$

9 $4\sqrt{18} \cdot 3\sqrt{2}$

10 $\sqrt{ab} \cdot \sqrt{ab}$

11 $\sqrt{3} \cdot 4\sqrt{2}$

12 $5\sqrt{3} \cdot 2\sqrt{3}$

13 $\sqrt{5} \cdot 6\sqrt{2}$

14 $6\sqrt{72}$

15 $\sqrt{11} \cdot \sqrt{44}$

16 $\sqrt{32} \cdot \sqrt{8}$

17 $\sqrt{10a} \cdot \sqrt{5a}$

18 $\sqrt{10r^2s} \cdot \sqrt{2r}$

19 $\left(4\sqrt{x^2y}\right) \cdot \left(3\sqrt{y}\right)$

20 $\sqrt{1n} \cdot \sqrt{1}$

Name _____ Date _____

Dividing Radicals

$$\frac{3\sqrt{13}}{4\sqrt{32}} \;=\; \frac{3\sqrt{13}}{4\sqrt{32}} \cdot \frac{\sqrt{2}}{\sqrt{2}} \;=\; \frac{3\sqrt{26}}{4\sqrt{64}} \;=\; \frac{3\sqrt{26}}{4 \cdot 8} \;=\; \frac{3\sqrt{26}}{32}$$

The process of expressing $\dfrac{3\sqrt{13}}{4\sqrt{32}}$ as $\dfrac{3\sqrt{26}}{32}$ is called rationalizing the denominator.

1 $\quad \dfrac{\sqrt{32}}{\sqrt{2}}$

2 $\quad \dfrac{\sqrt{45}}{\sqrt{5}}$

3 $\quad \dfrac{\sqrt{21}}{\sqrt{3}}$

4 $\quad \dfrac{\sqrt{5}}{\sqrt{15}}$

5 $\quad \dfrac{\sqrt{2}}{\sqrt{5}}$

6 $\quad \sqrt{\dfrac{3}{7}}$

7 $\quad \sqrt{\dfrac{19n^2}{32}}$

8 $\quad 3\sqrt{\dfrac{48}{9}}$

9 $\quad \sqrt{\dfrac{4b^2}{36}}$

10 $\quad \sqrt{\dfrac{9y^2}{25}}$

11 $\quad \dfrac{\sqrt{8r^2 s}}{\sqrt{2s}}$

12 $\quad 3\sqrt{\dfrac{3q}{2r}} \cdot \sqrt{\dfrac{q}{r}}$

Adding and Subtracting Radicals

$$7\sqrt{3} + 2\sqrt{48} \quad = \quad 7\sqrt{3} + 2\sqrt{16 \cdot 3} \quad = \quad 7\sqrt{3} + 2\left(4\sqrt{3}\right) \quad = \quad 7\sqrt{3} + 8\sqrt{3} \quad = \quad 15\sqrt{3}$$

Simplify.

1 $3\sqrt{5} + 7\sqrt{5}$

2 $7\sqrt{3} - 5\sqrt{3}$

3 $5\sqrt{2} + 3\sqrt{2}$

4 $8\sqrt{3} - 6\sqrt{3}$

5 $2\sqrt{8} + 5\sqrt{8}$

6 $4\sqrt{17} + 7\sqrt{17}$

7 $3\sqrt{5} + 8\sqrt{5} - 7\sqrt{5}$

8 $7\sqrt{17} + 2\sqrt{17} - 5\sqrt{17}$

9 $9\sqrt{7} + 2\sqrt{7} + 6\sqrt{7}$

10 $4\sqrt{13} - 2\sqrt{13} + 5\sqrt{13}$

11 $\sqrt{27} - \sqrt{3}$

12 $48\sqrt{3} - \sqrt{3}$

13 $2\sqrt{75} - \sqrt{3}$

14 $\sqrt{81} + \sqrt{36}$

15 $\sqrt{32} + 4\sqrt{8}$

16 $^-4\sqrt{75} + 3\sqrt{147}$

17 $6\sqrt{17} + 7\sqrt{6} - \sqrt{17}$

18 $\sqrt{338} - \sqrt{200} + \sqrt{162}$

19 $^-4\sqrt{2} - 8\sqrt{32} + 6\sqrt{72}$

20 $^-13\sqrt{5} + 17\sqrt{5}$

Algebra © 2004 Creative Teaching Press

The Pythagorean Theorem

How tall is the flag pole?

use $a^2 + b^2 = c^2$

$a^2 + 40^2 = 50^2$
$a^2 + 1600 = 2500$
$a^2 = 900$
$a = 30$

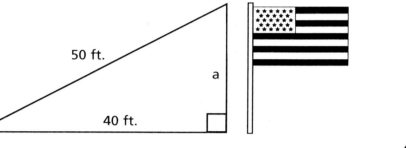

❶ What is the length of the hypotenuse?

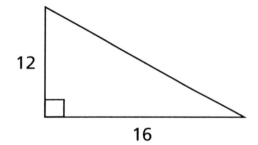

❷ Find the length of a diagonal of a rectangle whose dimensions are 56 in. by 33 in.

❸ Find the area of the square.

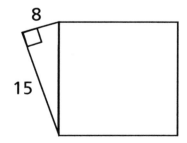

❹ A train travels 60 miles due west. It then travels due south 90 miles. How far is the train from its starting point?

Homework Help

You have helped Al improve his algebra skills. Yet there is more to be done. Find and correct his 4 mistakes.

Simplify.

1 $\sqrt{169} = 13$

2 $\sqrt{\dfrac{64}{100}} = \dfrac{8}{10}$

3 $\sqrt{529b^2} = 23b$

4 $\sqrt{8x} \cdot \sqrt{10x} = 16\sqrt{5}$

5 $\dfrac{\sqrt{33}}{\sqrt{3}} = \sqrt{11}$

6 $\sqrt{72} + \sqrt{8} = 10\sqrt{2}$

Using the Pythagorean Theorem for right triangles, fill in the blanks.

7 a = 9, b = 12, c = ___15___

8 a = ___20___, b = 21, c = 29

9 a = 12, b = ___35___, c = 37

10 a = ___18___, b = 24, c = 30

11 a = 15, b = ___33___, c = 39

12 a = 24, b = 32, c = ___40___

Quadratic Equations

$$(x+2)^2 = 81 \qquad \sqrt{x+2} = \sqrt{9} \qquad x+2 = {}^{\pm}9 \qquad x = 7 \text{ or } {}^{-}11$$

Solve. Express irrational solutions in simplest radical form.

1 $x^2 = 36$

2 $n^2 = 25$

3 $a^2 = 100$

4 $y^2 = 169$

5 $p^2 - 100 = 0$

6 $b^2 - 144 = 0$

7 $5d^2 = 180$

8 $9x^2 = 25$

9 $(a - 7)^2 = 0$

10 $b^2 + 3 = 19$

11 $z^2 - 48 = 0$

12 $4r^2 - 7 = 29$

13 $(a - 6)^2 = 36$

14 $(c + 4)^2 = 25$

15 $x^2 + 15 = 16$

16 $(z - 3)^2 = 32$

17 $y^2 = 7$

18 $j^2 = 40$

19 $3x^2 = 18$

20 $m^2 - 8 = 14$

21 $4g^2 = 32$

22 $3(2r + 7)^2 = 27$

23 $5e^2 - 13 = 47$

24 $x^2 = \dfrac{1}{81}$

The Quadratic Formula

$$x = \frac{-b \pm \sqrt{b^2 - 4ac}}{2a}$$

Hint: Remember standard form is $ax^2 + bx + c$.

$$\boxed{7x^2 - 6x + 1 = 0}$$

$$x = \frac{-(-6) \pm \sqrt{(-6)^2 - 4(7)(1)}}{2(7)}$$

$$x = \frac{6 \pm \sqrt{36 - 28}}{14}$$

$$x = \frac{6 \pm \sqrt{8}}{14}$$

$$x = \frac{6 \pm 2\sqrt{2}}{14}$$

$$x = \frac{3 \pm \sqrt{2}}{7}$$

Solve using the quadratic formula.

1 $x^2 - 9x + 14 = 0$

2 $a^2 - 2a - 8 = 0$

3 $3x^2 + 5x - 2 = 0$

7 $3k^2 + 3k - 20 = 0$

4 $2b^2 - 3b - 2 = 0$

8 $z^2 - 6z = 13$

5 $3y^2 + y - 10 = 0$

9 $6g^2 + g - 1 = 0$

6 $2x^2 - 6x - 8 = 0$

10 $p^2 = 4p$

More Practice with the Quadratic Formula

$$x = \frac{-b \pm \sqrt{b^2 - 4ac}}{2a}$$

Solve using the quadratic formula.

1 $x^2 - 10x + 22 = 0$

2 $x^2 + 10x + 18 = 0$

3 $2x^2 + 12x + 15 = 0$

4 $x^2 + 12x + 33 = 0$

5 $x^2 + 8x + 9 = 0$

6 $x^2 + 3x + 1 = 0$

7 $4x^2 - 4x - 1 = 0$

8 $x^2 + 4x + 1 = 0$

9 $3x^2 + 3x - 20 = 0$

10 $x^2 = {}^-3x + 2$

11 $x^2 = {}^-7x + 1$

12 $2x^2 + 4x + 1 = 0$

13 $x^2 = 5x + 4$

14 $x^2 = {}^-5x - 1$

15 $x^2 - 2x - 2 = 0$

16 $2x^2 + 14x + 23 = 0$

17 $x^2 = {}^-7x - 1$

18 $4x^2 - 32x + 61 = 0$

19 $x^2 - 2x - 8 = 0$

20 $4x^2 - 16x + 9 = 0$

Name _____ Date _____

Algebra Awareness (Quadratic Equations)

Solve. Express irrational solutions in simplest radical form.

1 $x^2 = 49$

   a) 7   b) ⁻7

   c) 0   d) ±7

2 $n^2 = 729$

   a) 29   b) ±27

   c) 27   d) ±29

3 What is the standard form of a linear equation?

   a) $a^2 + bx + c$ b) $ax^2 + by + c$

   c) $x^2 + bx + c$ d) $ax^2 + x + c$

4 $z^2 - 54 = 0$

   a) $2 \pm \sqrt{6}$ b) ± 2

   c) $3 \pm \sqrt{6}$ d) ± 3

5 $a^2 + 6a + 2 = 0$

   a) $^-3 \pm \sqrt{7}$ b) $3 \pm \sqrt{7}$

   c) $^-3 \pm \sqrt{6}$ d) $3 \pm \sqrt{6}$

6 $n^2 - 7n + 11 = 0$

   a) $\dfrac{^-7 \pm \sqrt{5}}{2}$ b) $\dfrac{7 \pm \sqrt{5}}{^-2}$

   c) $\dfrac{7 \pm \sqrt{5}}{2}$ d) $\dfrac{^-7 \pm \sqrt{5}}{^-2}$

7 $4z^2 - 36z + 79 = 0$

   a) $\dfrac{9 \pm \sqrt{2}}{^-2}$ b) $\dfrac{^-9 \pm \sqrt{2}}{2}$

   c) $\_\dfrac{9 \pm \sqrt{2}}{2}$ d) $\dfrac{9 \pm \sqrt{2}}{2}$

Answer Key

Real Number Line (page 5)

1.

2.
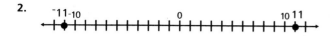

3.

4.
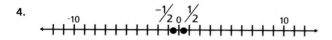

5. ALGEBRA IS FUN

Adding Integers (Real Numbers) (page 6)

1. ⁻5
2. 0
3. ⁻11
4. ⁻42
5. ⁻15
6. 9
7. ⁻6
8. ⁻9
9. 12.5
10. 29
11. 26.3
12. ⁻119
13. 308
14. 186
15. ⁻74.4
16. 20.7

17. positive
18. negative

Subtracting Integers (page 7)

1. ⁻3
2. 7
3. ⁻21
4. 26
5. ⁻5.2
6. 46
7. ⁻2
8. 6.5
9. ⁻63
10. ⁻12
11. 14.6
12. ⁻119
13. ⁻27
14. ⁻138
15. ⁻132
16. 132

17. positive
18. positive

Time for an Operation (page 8)

Correct problems are:

$2 - (⁻9) = 11$

$10 - (⁻3) = 13$ $⁻4 + 23 = 19$ $6 + (⁻12) = -6$

$⁻2 - (⁻1) = ⁻1$

Answer: addition

Multiplying Integers (page 9)

1. ⁻18
2. ⁻40
3. ⁻78
4. ⁻90
5. ⁻84
6. 18
7. ⁻9
8. ⁻40
9. 180
10. ⁻484
11. 0
12. ⁻4
13. 0
14. ⁻56

15. positive
16. negative
17. negative
18. positive

Dividing Integers (page 10)

1. ⁻2
2. 4
3. 12
4. 0
5. ⁻1.75
6. ⁻5
7. ⁻3
8. ⁻0.1875
9. 5
10. 0.3
11. ⁻21
12. 17
13. ⁻62
14. $1\frac{5}{21}$

15. addition, subtraction
16. negative
17. positive

Absolute Values (page 11)

1. 3
2. 42
3. 2
4. ⁻13
5. 18
6. ⁻17
7. 43
8. 6
9. 5
10. 5
11. 107
12. ⁻107
13. 0
14. 0
15. ⁻8
16. ⁻96

17. 0

Exponents (page 12)

1. 1,024
2. ⁻512
3. 25
4. 100
5. 8
6. 7,776
7. 1
8. 128
9. 81
10. 441
11. 7
12. ⁻1
13. 0
14. ⁻216

15. positive
16. negative

Homework Help (page 13)

2. ⁻7
3. 3.7
7. ⁻64.8

9. 51
11. ⁻39
12. ⁻8
14. 90
15. 36
16. -5

17. ⁻8, ⁻6, ⁻3, ⁻1, 2

Order of Operations (page 14)

1. ⁻27
2. 1
3. ⁻7
4. 20
5. 0
6. 13
7. 0

8. 11
9. 57
10. ⁻54
11. 27
12. 40
13. 396
14. 0

Algebra Awareness (Real Numbers) (page 15)

1. c
2. d
3. b
4. d

5. a
6. c
7. a
8. d

9. c

Always Treat the Substitute Properly (page 16)

1. 21
2. 28
3. 24
4. 34
5. 82
6. 70
7. 222
8. ⁻436

9. 126
10. 73
11. 30
12. 9
13. 8
14. ⁻54
15. ⁻26
16. 54

Combining Like Terms (page 17)

1. 10a
2. b
3. 3x + 6
4. 3x + 3y
5. ⁻6x + 3
6. 12x + 2
7. 2b + 4c
8. a − 3b
9. x + y
10. 9e − 2f

11. 5xy
12. ab
13. ⁻6xy + 15
14. ⁻10abc − 8
15. 2a + 3ef
16. 5uv + 6w
17. ⁻3xy − 9
18. 15x
19. $7y^2$
20. $2a^3b^2 − 4a^2b^3 + 2a^2$

Equations with Addition and Subtraction (page 18)

1. n = 2
2. n = ⁻18
3. n = 15
4. a = 7
5. b = 25.3
6. x = ⁻4 $\frac{3}{5}$
7. y = 87
8. n = ⁻67
9. x = 28
10. c = ⁻11

11. n = -2.5
12. n = ⁻7
13. b = 110
14. f = 97
15. x = 9.3
16. n = 7 $\frac{1}{2}$
17. c = 5
18. d = 27
19. f = 13
20. x = 7 $\frac{1}{3}$

Equations with Multiplication and Division (page 19)

1. a = 20
2. b = 2
3. f = 24
4. n = 40
5. b = 18
6. z = ⁻8
7. x = ⁻3
8. y = ⁻8

9. x = 12
10. n = ⁻35
11. x = 0
12. n = 2
13. a = ⁻2
14. v = ⁻7
15. n = ⁻2
16. b = 0.512

17. 5 meters

Solving Multi-Step Equations (page 20)

1. n = 32
2. b = ⁻2
3. f = 3
4. a = 1
5. x = ⁻6
6. y = 8
7. x = 9
8. z = 14

9. e = 15
10. a = 2
11. n = 8
12. b = 3
13. x = 6
14. n = 8
15. n = ⁻12
16. y = 9

An "Old" Question (page 21)

1. x = 2   B
2. n = 1   A
3. y = 2   B
4. b = 25   Y
5. c = 12   L
6. h = 15   O
7. a = 14   N

The Distributive Property (page 22)

1. $4n + 4$ **6.** $10z - 5$
2. $2a + 16$ **7.** $32 - 16k$
3. $^-70 - 49f$ **8.** $75 - 20a$
4. $10x + 15$ **9.** $6n + 7.2$
5. $3n - 27$ **10.** $5n - 41$

11. $n = 3$ **14.** $y = 2$
12. $x = {}^-8$ **15.** $a = 1$
13. $r = {}^-2$ **16.** $n = 2.2\overline{3}$

Variables on Both Sides (page 23)

1. $a = 2$ **10.** $c = {}^-9$
2. $x = 1$ **11.** $g = {}^-11$
3. $y = 3$ **12.** $b = 0$
4. $n = 8$ **13.** $x = 1$
5. $z = {}^-8$ **14.** $x = 9$
6. $f = {}^-6$ **15.** $u = 8$
7. $n = 12$ **16.** $h = {}^-2$
8. $d = 9$ **17.** $x = 4$
9. $j = 8$ **18.** $a = 0$

Translating Word Problems (page 24)

1. 8
2. 7
3. 73
4. 9
5. 60
6. $^-4$
7. 7
8. 80

What's the Word with Equations? (page 25)

1. pears = 20 cents   peaches = 10 cents
2. 1 7/8 of an hour or 1 hour and 52 1/2 minutes
3. 17
4. 2
5. 18, 73
6. 7

Consecutive Integers (page 26)

1. 9 and 10
2. $^-12$ and $^-11$
3. 24 and 25
4. 22, 23, and 24
5. $^-12$, $^-11$, $^-10$, and $^-9$
6. $^-28$, $^-27$, $^-26$, and $^-25$
7. 3 and 4
8. 84 and 86

Un-Doing Equations (page 27)

1. c
2. b
3. a
4. d $n = {}^-14$

Exponents and Variables (page 28)

1. x^4 **8.** $^-r^2$
2. $(^-y)^3$ **9.** d^2
3. a^5 **10.** u^3v^4
4. a^3b^2 **11.** $^-8k^2$
5. $10p^2$ **12.** $^-n^5$
6. $^-6n^2$ **13.** n^4
7. $^-54s^2$ **14.** $^-4a^2b^6$

15. 36   **17.** 35
16. $^-1$ **18.** $^-79$

Volume Challenge (page 29)

1. 36 ft³
2. 75 m³
3. 216 cm³
4. n cubed

Algebra Awareness (Equations) (page 30)

1. c
2. d
3. c
4. b
5. a
6. c
7. a
8. c

Add and Subtract Polynomials (page 31)

1. $6x^2 + 6$ **4.** $2s^2 - 2t + 2u$
2. $^-5x^3 + x$ **5.** $^-8a^4$
3. $^-n^5 + n + 2$ **6.** $^-z^2 + 2y^2 + 2x$

7. $^-12x^2 - 6x + 12$ **10.** $14x$
8. $13x^3$ **11.** 6
9. $9x^2 - x + 2$ **12.** $2a^2 - 3b^2 - 18b - 1$

Perimeter and Polynomials (page 32)

1. $6x + 4$
2. $24x^2 + 12$
3. $8x - 13$
4. $5x + 3$

Multiplying Polynomials by Monomials (page 33)

1. $9y^2 - 36$
2. $^-35b^2$
3. $12f^3 - 21f^2$
4. $^-4w^3 + 7w^2$
5. $^-5x^3 - 2x^2 + 3x$
6. $6a^3 - 15a^2 + 3a$

7. $2p^5 - 4p^4 + 16p^3$
8. $^-4m^2n^3 + 6m^3 + 16mn$
9. $11b^6 + 22b^4 + 55b^3$
10. $^-9a^6 + 7a^5 - 2a^4 + a^3$
11. $24x^5 - 8x^3y + 8x^2y^2$
12. $18n^5 + 6n^4 + 6n^2$

13. $x = 1$
14. $s = 2$

Multiplying Polynomials (page 34)

1. $2x^2 + 5x - 3$
2. $3r^2 - 11r - 20$
3. $j^2 - 9$
4. $4y^3 + 2y^2 + 8y + 4$
5. $8n^3 - 72n^2 - 6n + 54$
6. $x^3 + x^2 - 27x - 35$
7. $2b^3 - 3b^2 + 4b - 3$

8. $p^3 + 3p^2 - 2p + 40$
9. $54y^3 - 15y^2 + 37y - 4$
10. $p^3 + 3p^2 - 2p + 40$
11. $3n^3 - 35n^2 + 58n - 80$
12. $5s^3 - 17s^2 + 10s - 12$
13. $5x^3 - x^2 + 5x - 9$
14. $4r^3 - 37r^2 - 85r + 88$

Distance, Rate, and Time (page 35)

1. 5 hours 6 minutes
2. 10:11 a.m.
3. 2:20 a.m.
4. 3 hours 10 minutes

Homework Help (page 36)

2. x^4
4. $^-2g + 6$
7. $z^3 + 3z^2 - z$

10. $15x^2 - 22x - 9$
12. $36d^2 + 48d + 16$
14. $25x^2 - 60x + 36$
16. $x^2 - 2xy + y^2$

Area and Polynomials (page 37)

1. $2x^2 - 2x$
2. $25x^2 - 40x + 16$
3. $x^2 - 9/2x - 28$
4. $9x^2/2$

Area and Perimeter Challenge (page 38)

1. $12x^2 + 14x - 8$
2. $3x^2 + 5x + 2$
3. $10x^2 + 11x + 3$

FOILed (page 39)

1. $x^2 + 2x - 3$
2. $x^2 + x - 20$
3. $6x^2 - 4x - 16$
4. $63x^2 + 8x - 16$
5. $36x^2 + 6x - 2$

6. $5x^2 - x - 4$
7. $14x^2 - 25x + 6$
8. $2x^4 - 8x^2 - 24$
9. $5x^4 - 52x^2 + 63$
10. $9x^4 + 18x^2 + 9$

11. No, because the product of the outside terms cancel out with the product of the inside terms.
12. $4x^4 - 9$

Squaring Binomials (page 40)

1. $x^2 - 2x + 1$
2. $9x^2 + 6x + 1$
3. $9x^2 + 24x + 16$
4. $x^2 - 8x + 16$
5. $4x^2 - 8x + 4$

6. $81x^2 + 72x + 16$
7. $4x^2 - 28x + 49$
8. $x^4 + 12x^2 + 36$
9. $16x^6 - 16x^3 + 4$
10. $64x^8 - 48x^4 + 9$

11. $2x + 2$
12. $3x - 1$

Polynomial Equations (page 41)

1. $a = 4$
2. $v = ^-7$
3. $c = 6$
4. $n = 14$
5. $b = 2$
6. $x = ^-3$
7. $a = 4$

8. $y = 10$
9. $x = 13$
10. $n = ^-5/2$
11. $a = 20/11$
12. $x = 5$
13. $x^4 - 8x^3 + 26x^2 - 40x + 25$
14. $x^3 + 15x^2 + 75x + 125$

What's the Word with Rectangles? (page 42)

1. 273 ft^2
2. W = 24 m L = 72 m
3. L = 9 ft   W = 4.5 ft
4. W = 67 in L = 54 in

Algebra Awareness (Polynomials) (page 43)

1. b
2. c
3. d
4. d
5. a
6. b
7. a
8. d

What's the Word with More Equations? (page 44)

1. 10
2. 4,926
3. 7:50 p.m.
4. $2,500

Common Factors (page 45)

1. 14
2. 8
3. 1
4. 21
5. 33
6. 7
7. 84

8. 11
9. 20
10. 4
11. 4
12. 2
13. 4
14. 12

Greatest Monomial Factor (page 46)

1. 4
2. 12n
3. y
4. 5c

5. x^3
6. 17
7. 3
8. 1

9. $4x^2y$
10. $3m^2n^5$
11. $^-6ab^3$

12. $3c^6d$
13. $^-6x^5$
14. $^-12s^4t$

Prime Time (page 47)

The boxes with the following terms are shaded:

24, 32 **8**

2m, 4 **2**

10n, 15n **5n**

20a, 32a **4a**

44x, 22x **22x**

2xy, 7xy **xy**

8, 12, 24 **4**

4z, 6z, 8z **2z**

6b, 8b, 24b **2b**

Answer: 2

Difference of Squares (page 48)

1. (x + 1) (x − 1)
2. (s − 4) (s + 4)
3. (v + 6) (v − 6)
4. (n − 9) (n + 9)
5. (2d + 2) (2d − 2)
6. $(5n^2 − 4) (5n^2 + 4)$
7. (4n + 7) (4n − 7)

8. (m − 5) (m + 5)
9. (3a + 10) (3a − 10)
10. $(25x^2 − 1) (25x^2 + 1)$
11. $(s^2 + 9v^2) (s^2 − 9v^2)$
12. $(x^4 − y^2) (x^4 + y^2)$
13. $(9m^2 + 4n^2) (9m^2 − 4n^2)$
14. $(1 − 11c^8) (1 + 11c^8)$

Circular Spaces (page 49)

1. $2r^2 (4 − \pi)$
2. $r^2 (8 + \pi)$
3. $4r^2 (4 − \pi)$
4. $r^2 (4 − \pi)$

Factoring Perfect Square Trinomials (page 50)

1. $z^2 − 2z + 1$
2. $n^2 + 10n + 25$
3. $25k^2 + 30k + 9$

4. $9x^2 − 12xy + 4y^2$
5. $144j^2 − 240jk + 100k^2$
6. $36m^2 − 12mn + n^2$

7. not a perfect square
8. $(2y + 1)^2$
9. not a perfect square

10. $(a + 1)^2$
11. not a perfect square
12. $(f^2 − 7)^2$

13. They are positive.
14. It is the same sign as the middle term.

Factoring Pattern for $x^2 + bx + c$ (page 51)

1. (x + 4) (x + 1)
2. (a + 4) (a + 3)
3. (n + 3) (n + 1)
4. (s − 4) (s − 2)
5. (g + 2) (g + 7)
6. (d − 8) (d − 2)
7. (b − 9) (b − 4)
8. (v + 5) (v + 3)
9. (x − 13) (x − 2)

10. (a − 3) (a − 5)
11. (x + 4) (x + 1)
12. (p + 6) (p + 4)
13. (j − 4) (j − 2)
14. (x + 5) (x + 2)
15. (16 − b) (4 − b)
16. (3 − r) (1 − r)
17. (6 − q) (5 − q)
18. (13 − y) (2 − y)

Factoring Pattern for $x^2 + bx − c$ (page 52)

1. (x + 6) (x − 1)
2. (a − 6) (a + 2)
3. (n + 4) (n − 2)
4. (s − 4) (s + 1)
5. (g − 8) (g + 6)
6. (d − 9) (d +5)
7. (b − 15) (b + 2)
8. (v + 4) (v − 2)
9. (x + 9) (x − 2)

10. (a − 8) (a + 4)
11. (x − 6) (x + 4)
12. (p + 7) (p − 3)
13. (j − 27) (j + 2)
14. (x − 5) (x + 3)
15. (s + 8) (s − 4)
16. (f + 10) (f − 4)
17. (x + 8y) (x − 7y)
18. (r − 16s) (r + 3s)

Factoring Pattern for $ax^2 + bx + c$ (page 53)

1. (5n − 3) (n − 4)
2. (3a + 1) (a + 2)
3. (2z − 1) (z − 7)
4. (7b + 5) (b + 8)
5. (5y − 1) (y + 7)
6. (5y − 3) (y − 4)
7. (5f − 1) (f + 1)
8. (10c + 1) (c − 6)
9. (5g + 6) (g + 12)

10. prime
11. (3p − 2) (p + 3)
12. (2c + 3) (2c − 1)
13. (2r − 1) (r − 4)
14. (3q + 4) (q − 1)
15. (10b + 3) (b − 2)
16. (6s − t) (s − 3t)
17. (3e + 2f) (e + 3f)
18. (6a + 5) (a + 2)

Factoring by Grouping (page 54)

1. $(2 + x)(x - 3)$
2. $(5 + b)(a + 3)$
3. $(3 + s)(r - 4)$
4. $(6 + b)(a + 7)$
5. $(e - 4)(f - g)$
6. $(4 + x)(z + 3)$
7. $(s - 5)(t - 3)$
8. $(m - 8)(n - 4)$
9. $(3a - 2)(6a - b)$
10. $(r + 7)(q - 1)$

11. $(3y + 4)(x + 4)$
12. $(r + t)(3 + s)$
13. $(a + c)(b + 5)$
14. $^-7x^2(x - 3) - 10(x - 3)$
15. $(6s^2 - 5)(4s - 1)$
16. $(a^2 + 3)(a - 2)$

Factoring Checklist (page 55)

A.
1. no
2. yes $(x + 11)(x - 11)$

B.
1. no
2. no
3. no
4. yes $(7x + 5)(x + 8)$

C.
1. no
2. no
3. no
4. no
5. yes $(4n^2 - 3)(5n + 6)$

Solving Equations by Factoring (page 56)

1. $^-2$ or 5
2. 1 or 7
3. $^-9$ or $^-1$
4. 3 or 9
5. $^-5$ or $^-4$
6. $^-2$ or 6
7. 0 or 3/4
8. $^-\frac{5}{2}$, $^-\frac{1}{2}$, or 0
9. 1
10. $^-\frac{1}{4}$ or 4
11. $^-\frac{5}{6}$ or 0
12. $^-\frac{1}{6}$ or $\frac{1}{6}$
13. $^-10$ or 10
14. 0 or 16
15. $^-1$ or $\frac{2}{3}$
16. $1\frac{1}{2}$ or $5\frac{1}{2}$

Algebra Awareness (Factoring) (page 57)

1. c
2. a

3. d   6. c
   a   7. d
              8. d

Simplifying Algebraic Fractions (page 58)

1. x
2. $2a^2$
3. $5r^2$
4. $12d^3$
5. 24b
6. y
7. 1
8. $3u^3$
9. a/2
10. $^{12}\!/_{x^2y^2}$
11. $12t^4$
12. $3/m^3n$
13. $^{-2e^4}\!/_f$
14. $^{15}\!/_{r^3s^4}$
15. $^{ab^2}\!/_3$
16. $^{-9y^4}\!/_{16x^4}$

Dividing Polynomials by Monomials (page 59)

1. $2n + 3$
2. $4y - 2$
3. $6x - 3$
4. $2r - 1$
5. $3f + 1$
6. 10
7. $f - 3$
8. $\frac{2a + b}{a} + \frac{b}{a}$
9. $3x^3 - x^2 - 4x$
10. $^-4w^2 + 2w + 3$
11. $a^4b^3 - a^3b^2 - a^2b$
12. $4m^5n^4 - 4m^2n - m$

More Algebraic Fractions (page 60)

1. $\dfrac{x + y}{2}$
2. 5
3. $\dfrac{6}{b - 5}$
4. $\dfrac{r - 2}{r}$
5. $\dfrac{7}{c + d}$
6. $\dfrac{y + 4}{y - 4}$
7. $7\dfrac{5 - f}{f + 7}$
8. $\dfrac{6s^2}{s + 5}$
9. $\dfrac{5x + 3y}{10x^2 + 6y^2}$
10. $\dfrac{2w - 3}{2}$
11. $\dfrac{a}{a - 5}$
12. $\dfrac{b - 2}{2b}$
13. $-\dfrac{2n - 1}{2n}$
14. $\dfrac{3x - 7}{x - y}$

Multiplying Fractions (page 61)

1. $5/3$

2. $-1/25$

3. a/c

4. $6x/y^2$

5. $\dfrac{4}{n+1}$

6. $\dfrac{x}{x-2}$

7. $a^2 + a$

8. $\dfrac{2a + 2c}{a}$

9. $\dfrac{(k-8)(k-5)}{(k+7)^2}$

10. $\dfrac{x^2 + y^2}{2x + 2y}$

11. $-\dfrac{2}{5y - 2x}$

12. $4ab - 2b^2$

13. $\dfrac{5a}{(a+b)^2}$

14. $\dfrac{1}{ab}$

Area with Fractions (page 62)

1. $\dfrac{9x^2}{49}$

2. $\dfrac{x^2 - 36}{4}$

3. $1/3$

4. $\dfrac{8x^2}{9}$

Dividing Fractions (page 63)

1. $2/5$

2. $6/y$

3. $a^3/8$

4. $-9/4$

5. $\dfrac{e+2}{e+3}$

6. $\dfrac{3}{x+1}$

7. $\dfrac{n-2}{2n}$

8. $\dfrac{6}{a^2 - ab}$

9. 1

10. $x - y$

11. $\dfrac{1}{a^2 + a - 6}$

12. $\dfrac{-3}{2r^2 + 2r - 12}$

13. $\dfrac{6y}{x^3}$

Mix It Up! (page 64)

1. $\dfrac{5x - 10}{x - 4}$

2. 1

3. $\dfrac{1}{2 + y}$

4. $\dfrac{2a + 6}{a - 2}$

5. $\dfrac{1}{4m - 4n}$

6. $\dfrac{3}{r + 3s}$

Homework Help (page 65)

3. $\dfrac{4s^2 - 5s - 6}{2s^2 + 4s}$

6. $\dfrac{28e^2 + 4ef}{7e^3 + 7e^2f + ef + f^2}$

7. $\dfrac{6n^2}{n + 5}$

11. $\dfrac{16}{m - n}$

Least Common Denominators (page 66)

1. 10

2. 6

3. xy

4. xy^2

5. 14a

6. xy

7. $x^2 + 1$

8. $12a^2$

9. $8a^2b^2$

10. $8r^2s^2t$

Adding and Subtracting Fractions with Like Denominators (page 67)

1. $5x/9$

2. $\dfrac{6r + 8s}{4b}$

3. 1

4. $\dfrac{x^2 + 4}{x + 2}$

5. $\dfrac{5}{6a^2bc}$

6. $\dfrac{f}{2}$

7. $\dfrac{s - 4}{s - 3}$

8. $\dfrac{5y + 7}{3}$

9. 1

10. $q + 1$

Fill in the Blanket (page 68)

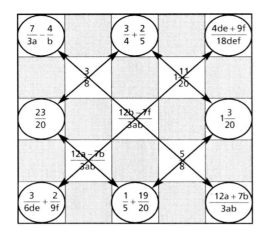

$$\frac{7}{3a} - \frac{4}{b}$$ $$\frac{3}{4} + \frac{2}{5}$$ $$\frac{4de + 9f}{18def}$$

$$\frac{3}{8}$$ $$1\frac{11}{20}$$

$$\frac{23}{20}$$ $$\frac{12b - 7f}{3ab}$$ $$1\frac{3}{20}$$

$$\frac{12a - 7b}{3ab}$$ $$\frac{5}{8}$$

$$\frac{3}{6de} + \frac{2}{9f}$$ $$\frac{1}{5} + \frac{19}{20}$$ $$\frac{12a + 7b}{3ab}$$

Adding and Subtracting Fractions with Unlike Denominators (page 69)

1. $\dfrac{4x + 3}{2y}$

2. $\dfrac{15a}{15b}$

3. $\dfrac{^-2z + 2}{w - 2}$

4. $\dfrac{6 + 4x}{3 - 2x}$

5. $\dfrac{2ry + 1}{y^2}$

6. $\dfrac{n^2 + 7n - 5}{5n^2}$

7. $\dfrac{^-k + 2}{2k - 2}$

8. $\dfrac{^-8e - 17f}{30}$

9. $\dfrac{a^2 + 2a + 8}{4a^2}$

10. $\dfrac{n + 23}{24}$

Perimeter and Fractions (page 70)

1. $\dfrac{7z + 3}{6}$

2. $3\dfrac{2}{3}x$

3. $\dfrac{14a}{3}$

Mixed Expressions (page 71)

1. $21/8$

2. $67/7$

3. $\dfrac{3a + 5}{a}$

4. $\dfrac{3x^2 - 2}{x}$

$\dfrac{6h - 5}{^-3}$

6. $\dfrac{10n + 4}{n}$

7. $\dfrac{f^2 - 4f + 7}{f - 2}$

8. $\dfrac{6z^2 - z - 6}{3z - 1}$

9. $\dfrac{b^3 - b^2 - b - 1}{b - 1}$

10. $\dfrac{2x^2 - 8y^2}{2x - 3y}$

Algebra Awareness (Fractions) (page 72)

1. c
2. b
3. d
4. d
5. b

What's the Word with Age? (page 73)

1. 12
2. Larry = 21, Moe = 7, Curly = 5
3. Steve = 52, Betty = 39
4. $7\frac{1}{6}$
5. 40

Ratios (page 74)

1. 6:19
2. 5:24
3. 1:16
4. 1:3
5. 3:16
6. 1:2
7. 7:11
8. 13:8
9. 2

What's the Word with Ratios? (page 75)

1. 18, 32
2. 45, 60
3. 162
4. L = 20 W = 5
5. boys = 14 girls = 21
6. 360
7. Diem = 30 miles per hour   Kevin = 12 miles per hour

Proportions (page 76)

1. n = 16
2. m = 4
3. $p = 28/15$
4. r = 1
5. $q = 26/39$
6. u = 9
7. a = 3
8. w = 125
9. $a = ^-5/2$
10. n = 2
11. b = 0.9
12. y = 1

What's the Word with Percents? (page 77)

1. 20%
2. 25%
3. 64%
4. 2%
5. 30
6. 90
7. 42
8. 1,080
9. 400
10. 18
11. $1,320

Equations with Fractional Coefficients (page 78)

1. $x = 6$
2. $e = 6$
3. $y = 4$
4. $k = \frac{7}{11}$
5. $z = 12$
6. $p = 36$
7. $b = 0$
8. $y = 3/2$ or $1\frac{1}{2}$
9. $s = 6$
10. $n = 0$
11. $v = 16$
12. $d = \frac{-14}{19}$

Fractional Equations (page 79)

1. $x = 1$
2. $m = 3$
3. $r = 6$
4. $g = 5$
5. $s = 5$
6. $y = \frac{-1}{9}$
7. $x = 8$
8. $n = {}^-8$
9. $b = 2$
10. no solution
11. $x = \frac{-9}{4}$ or $^-2\frac{1}{4}$
12. $a = \frac{-1}{2}$ or 1

Mixture Problems (page 80)

1. one liter
2. 25 gallons
3. 11 pennies, 55 nickels, 20 dimes, 14 quarters
4. $12.00 per hour
5. 75 ounces

Homework Help (page 81)

2. Length = 21; Width = 7
3. $r = 3$
5. $n = {}^-4$

Equations in Two Variables (page 82)

1. yes   no
2. no   yes
3. no   no
4. no   no
5. no   yes
6. yes   no
7. yes   yes
8. no   no
9. 5   0
10. 0 $^-1$
11. 0.5 $^-1$
12. 4   11
13. 10   14
14. $^-5$ $^-4$
15. 0   4
16. $^-4$ $\frac{-10}{3}$

Graphing Coordinates (page 83)

I.
1. ($^-4$, $^-6$)
2. (3, $^-8$)
3. (0, 2)
4. ($^-7$, 2)
5. (1, $^-5$)
6. (3, 6)

II.
7. H
8. D
9. J
10. F
11. B
12. L

13. counter-clockwise

III.
14. y-axis
15. quadrant II
16. quadrant III
17. quadrant I
18. x-axis

Slope of a Line (page 84)

1. 2
2. $^-3/8$
3. 1/4
4. 2
5. $^-3/2$
6. 2/3
7. 5/3
8. 5/2

Slope Sleuth (page 85)

Cartesian

The Slope-Intercept Form (page 86)

1. $y = {}^-5x - 3$ ${}^-5, 3$

2. $y = {}^-3x + 8$ ${}^-3, 8$

3. $y = x - 5$ $1, {}^-5$

4. $y = \frac{4}{3}x + \frac{7}{3}$ $\frac{4}{3}, \frac{7}{3}$

5. $y = 3x + 1$ $3, 1$

6. $y = \frac{1}{2}x + 2$ $\frac{1}{2}, 2$

7. $y = \frac{4}{3}x - 3$

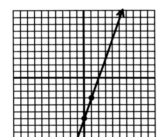

8. $y = \frac{1}{5}x + 1$

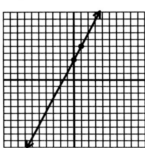

9. $y = 3x - 6$

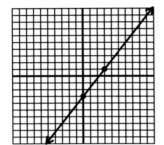

10. $y = 2x + 3$

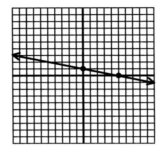

More Practice with the Slope-Intercept Form (page 87)

1. $y = {}^-2x + 5$

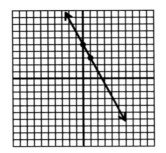

2. $y = x - 4$

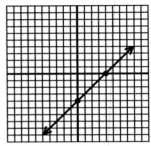

3. $y = -\frac{1}{3}x + 2$

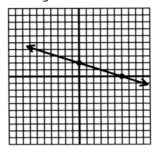

4. $x = 4$

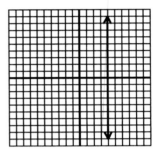

5. $y = {}^-2x$

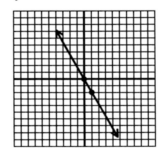

6. $y = x - 6$

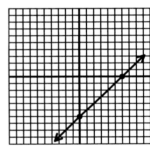

Algebra Awareness (Graphing) (page 88)

1. d   **2.** b

3. d   **4.** a

5. d   **6.** d

7. d   **8.** d

Translating Word Problems with Multiple Unknowns (page 89)

1. 19

2. 61

3. 73, 15

4. 81, 21

5. 542, 193

6. 180, 29

Systems of Linear Equations: The Graphing Method (page 90)

1. (3, 3)

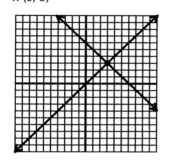

2. (2, ⁻2)

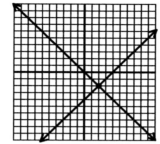

3. (3, ⁻2)

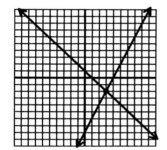

4. (2, ⁻1)

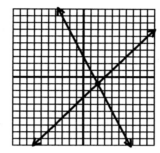

5. (2, 0)

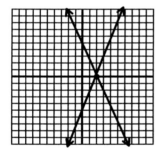

6. (⁻1, ⁻3)

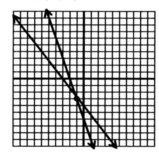

Systems of Equations: The Substitution Method (page 91)

1. (4, 24)   **8.** (0, 1)

2. (6, 18)   **9.** (8, 1)

3. (10, 2½)   **10.** (⁻5, 8)

4. (⁻10, 10)   **11.** (⁻1, ⁻10)

5. (0, 5)   **12.** (4, 2)

6. (2, 4)   **13.** (0, 5)

7. (4, 6)   **14.** (5, 10)

Systems of Equations: The + or – Method (page 92)

1. (5, 2)   **8.** (0, 1)

2. (6, ⁻1)   **9.** (0, 3)

3. (2, ⁻1)   **10.** (1, 0)

4. (7, 1)   **11.** (0, 1)

5. (4, 0)   **12.** (6, ⁻1)

6. (3, 3)   **13.** (2, 1)

7. (8, ⁻8)   **14.** (7, 9)

Systems of Equations: Multiplication with the + or – Method (page 93)

1. (2, 1)   **8.** (⁻7, 3)

2. (⁻59, 36)   **9.** (⁻7, 4)

3. (3, 1)   **10.** (⁻8, 0)

4. (0, 0)   **11.** (2, 2)

5. (⁻8, ⁻7)   **12.** (1, 3)

6. (⁻8, 6)   **13.** (7, 0)

7. (⁻6, 4)   **14.** (5, 9)

Puzzle Problems (page 94)

1. Denise = 16   sister = 10

2. Cathy = 18   Kitty = 12

3. 71

4. 96

5. 2/9

6. 4/5

Homework Help (page 95)

2. (4, 2)

6. Hank = 41   Bethany = 9

Inequalities (page 96)

1.

2.

3.

4.

5. T   **8.** T
6. F   **9.** T
7. T   **10.** T

Solving Inequalities (page 97)

1. $n \geq 8$
2. $y > 4$
3. $p < 2$
4. $z \leq 8$
5. $x < {}^-20$
6. $b > 3$
7. $k \leq {}^-2$
8. $m < {}^-3$

9. $d < 0$

10. $w \leq {}^-4$

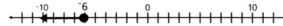

11. $x \leq {}^-6$

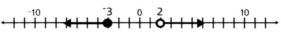

Combined Inequalities (page 98)

1.
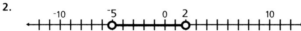

2.
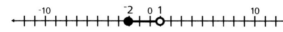

3.
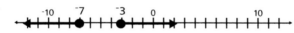

4.

5.
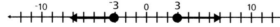

Absolute Value and Open Sentences (page 99)

1.
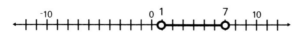

2.
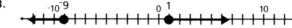

3.
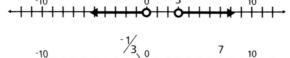

4.

Graphing Linear Inequalities (page 100)

1.
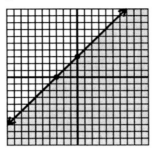

2.
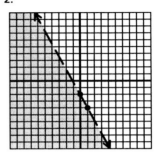

3.
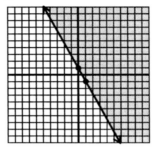

4.
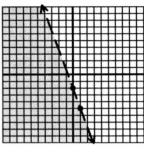

Systems of Linear Inequalities (page 101)

1.
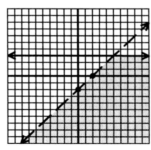

2.
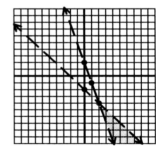

3.
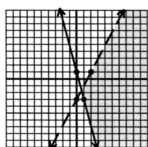

4.
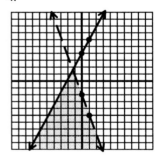

Algebra Awareness (Inequalities) (page 102)

1. d
2. c
3. a
4. b
5. d

Radicals (page 103)

1. 4	9. ±7
2. 7	10. ±13
3. 4	11. ±35
4. 121	12. ± $\frac{1}{3}$
5. ⁻43	13. ± $\frac{1}{16}$
6. $\frac{1}{8}$	14. ⁻3
7. ⁻14	15. $\frac{11}{5}$
8. ⁻ $\frac{2}{3}$	16. ⁻ $\frac{2}{3}$

Is That Rational? (page 104)

3	$\sqrt{64}$	0.79	$\sqrt{25}$	0.79	$\sqrt{1.44}$	2
$\sqrt{\frac{28}{63}}$	$(\sqrt{2})^4$	7.3	$\sqrt{16}$	2	$\sqrt{9}$	2
0.123	$\sqrt{\frac{243}{147}}$	$6.7\overline{8}$	5	19.1	$\sqrt{50}$	$7-\sqrt{11}$
$\sqrt{3}$	$\sqrt{225}$	4	$\sqrt{36}$	$5+\sqrt{7}$	2.9	$\sqrt{7}-\sqrt{7}$
11	7π	$\sqrt{100}$	1.23...	0.9	13	$\sqrt{4}$
$\sqrt{1.69}$	23	$\sqrt{2}$	$\sqrt{\frac{1}{36}}$	2	2.91	$\sqrt{\frac{4}{25}}$
5.23	$\sqrt{7 \cdot 7}$	3.4	2.1	$\sqrt{3}-\sqrt{3}$	17	$\sqrt{9 \cdot 9}$

Square Roots with Variables (page 105)

1. $5\|x\|$	10. $7\|r^3 s^3\|$
2. $11\|b\|$	11. $5\|q\|\sqrt{3q}$
3. $9\|n\|$	12. $\pm 2\|a^3\|b^2\sqrt{14}$
4. $6z^2$	13. $\frac{\|f\|}{7}$
5. $8\|rs\|$	14. $\frac{e^2 f^2}{13}$
6. $10m^2 n^2$	15. $\frac{3}{z^2}$
7. ⁻$3c^2$	16. $\pm 3\|mn\|\sqrt{6n}$
8. $4\|bc\|\sqrt{5}$	17. $4\|v^3\|\sqrt{5}$
9. $12\|h\|$	18. $\pm\frac{5\|y^5\|}{4}$

Multiplying Radicals (page 106)

1. $\sqrt{10}$	11. $4\sqrt{6}$
2. $\sqrt{18}$	12. 30
3. 2	13. $6\sqrt{10}$
4. 6	14. $36\sqrt{2}$
5. 8	15. 22
6. 12	16. 16
7. 10	17. $5a\sqrt{2}$
8. 6	18. $2r\sqrt{5rs}$
9. 72	19. 12xy
10. ab	20. $\sqrt{n}$

Dividing Radicals (page 107)

1. 4	7. $\frac{n\sqrt{38}}{8}$
2. 3	8. $4\sqrt{3}$
3. $\sqrt{7}$	9. $\frac{b}{3}$
4. $\frac{\sqrt{3}}{3}$	10. $\frac{3y}{5}$
5. $\frac{\sqrt{10}}{5}$	11. 2r
6. $\frac{\sqrt{21}}{7}$	12. $\frac{3q\sqrt{6}}{2r}$

Adding and Subtracting Radicals (page 108)

1. $10\sqrt{5}$	11. $2\sqrt{3}$
2. $2\sqrt{3}$	12. $47\sqrt{3}$
3. $8\sqrt{2}$	13. $9\sqrt{3}$
4. $2\sqrt{3}$	14. 15
5. $14\sqrt{2}$	15. $12\sqrt{2}$
6. $11\sqrt{17}$	16. $\sqrt{3}$
7. $4\sqrt{5}$	17. $7\sqrt{6} + 5\sqrt{17}$
8. $4\sqrt{17}$	18. $-14\sqrt{2}$
9. $17\sqrt{7}$	19. 0
10. $7\sqrt{13}$	20. $4\sqrt{5}$

The Pythagorean Theory (page 109)

1. 20
2. 65 in.
3. 289 units^2
4. $30\sqrt{13}$

Homework Help (page 110)

2. $\dfrac{4}{5}$
4. $4x\sqrt{5}$
6. $8\sqrt{2}$
11. 36

Quadratic Equations (page 111)

1. ±6
2. ±5
3. ±10
4. ±13
5. ±10
6. ±12
7. ±6
8. $\pm\dfrac{5}{3}$
9. 7
10. ±4
11. $\pm4\sqrt{3}$
12. ±3

13. 0 and 12
14. ⁻9 and 1
15. ±1
16. $3\pm4\sqrt{2}$
17. $\pm\sqrt{7}$
18. $\pm2\sqrt{10}$
19. $\pm\sqrt{6}$
20. $\pm\sqrt{22}$
21. $\pm2\sqrt{2}$
22. ⁻5 and ⁻2
23. $\pm2\sqrt{3}$
24. $\pm\dfrac{1}{9}$

The Quadratic Formula (page 112)

1. 2 and 7
2. ⁻2 and 4
3. ⁻2, 3, and 5
4. ⁻$\dfrac{1}{2}$ and 2
    ⁻2 and $\dfrac{5}{3}$

6. ⁻1 and 4
7. $\dfrac{3\pm\sqrt{249}}{6}$
8. $3\pm\sqrt{22}$
9. ⁻$\dfrac{1}{2}$ and $\dfrac{1}{3}$
10. 0 and 4

More Practice with the Quadratic Formula (page 113)

1. $5\pm\sqrt{3}$
2. ⁻$5\pm\sqrt{7}$
3. $\dfrac{⁻6\pm\sqrt{6}}{2}$
4. ⁻$6\pm\sqrt{3}$
5. ⁻$4\pm\sqrt{7}$
6. $\dfrac{⁻3\pm\sqrt{5}}{2}$
7. $\dfrac{1\pm\sqrt{2}}{2}$
8. ⁻$2\pm\sqrt{3}$
9. $\dfrac{⁻3\pm\sqrt{249}}{6}$
10. $\dfrac{⁻3\pm\sqrt{17}}{2}$

11. $\dfrac{⁻7\pm\sqrt{53}}{2}$
12. $\dfrac{⁻2\pm\sqrt{2}}{2}$
13. $\dfrac{5\pm\sqrt{41}}{2}$
14. $\dfrac{⁻5\pm\sqrt{21}}{2}$
15. $1\pm\sqrt{3}$
16. $\dfrac{⁻7\pm\sqrt{3}}{2}$
17. $\dfrac{⁻7\pm\sqrt{45}}{2}$
18. $\dfrac{8\pm\sqrt{3}}{2}$
19. ⁻2 and 4
20. $\dfrac{4\pm\sqrt{7}}{2}$

Algebra Awareness (Quadratic Equations) (page 114)

1. d
2. b
3. b
4. c
5. a
6. c
7. d